Craig Madden has years  in domestic water manage— more than 500 greywater and rainwater systems for households throughout Victoria and New South Wales. He runs a business called 'The Water Bloke', providing water-saving solutions to a rapidly growing market. Visit him at www.thewaterbloke.com.au.

Amy Carmichael is a journalist with a passion for helping people understand complex environmental issues. Her front-page articles about the water shortage in her hometown of Vancouver were read around the world. She is now based in Australia.

EVERY LAST DROP

Craig Madden
&
Amy Carmichael

RANDOM HOUSE AUSTRALIA

Random House Australia Pty Ltd
100 Pacific Highway, North Sydney NSW 2060
www.randomhouse.com.au

Sydney New York Toronto
London Auckland Johannesburg

First published by Random House Australia 2007

Copyright © Craig Madden and Amy Carmichael 2007

All rights reserved. No part of this publication may be reproduced, stored in a retrieval system, or transmitted in any form or by any means, electronic, mechanical, photocopying, recording or otherwise, without the prior written permission of the publisher.

National Library of Australia
Cataloguing-in-Publication Entry

Madden, Craig (Craig Matthew), 1974–.
Every last drop: the water saving guide.

Bibliography.
ISBN 978 1 74166 888 9 (pbk.).
    1. Water conservation – Australia. I. Carmichael, Amy
    (Amy McKenzie), 1979–.
    II. Title.

333.91160994

Cover illustration by Photolibrary
Cover design by Darian Causby, Highway 51 Design Works
Internal design and typesetting by VJ Battersby
Internal images © Getty Images, Newspix, Craig Madden and Amy Carmichael
Printed and bound by Griffin Press, South Australia

10 9 8 7 6 5 4 3 2 1

Extract of John Marsden from *7:30 Report*, 'Howard pledges $10b to solve water crisis', reporter Matt Peacock, first broadcast on Saturday 25 January 2007, is reproduced by permission of the Australian Broadcasting Corporation and ABC Online. © 2007 ABC. All rights reserved.

To the memory of Kristian, a true friend.
—Craig

To Ted, Kim and Sandy, your love inspires me.
— Amy

CONTENTS

Craig's story — ix
Amy's story — xiii
Introduction — xv

Part 1 – Water: the facts
The problem so far — 3
How much water do we use? — 17

Part 2 – Saving water at home
Around the house — 27
Kitchen — 41
Bathroom — 57
Toilet — 81
Laundry — 99
Garden — 113
Pools, water pistols and running through the sprinkler — 145
Garage and driveway — 151
Rainwater harvesting — 159
Recycling greywater at home — 183

Part 3 – The big picture

Where does our water come from?	213
The Murray–Darling Basin	223
Dammed if they do	229
Our leaky infrastructure	239
Stormwater reuse	243
Water recycling projects around Australia	255
Desalination: the technosolution?	265
Agriculture: the great debate	281
Risky business	299
Water politics	313
Education	323
The future: ours to hold!	333

Part 4 – Resources

Water conservation programs	341
Broader conservation and sustainability programs	347
Craig's recommended products	350
Amy's recommended products	356
Water authorities	360
Green power suppliers	366
Rebates	368

Acknowledgements	377

CRAIG'S STORY

As a young child I lived in the country town of Hamilton, Victoria, 'Sheep Capital of the World.' Although my parents were never farmers, we had a rural lifestyle. My parents owned the local nursery and florist, so I guess it's fair to say I've always had an interest in plants and irrigation, and therefore water. My father taught us how to pot plants and arrange flower wreaths, and my brother and I got used to potting for pocket money every weekend.

In the early eighties we moved to the Sunshine Coast in Queensland, but Dad kept working in the nursery industry. As teenagers, my brother and I made quite a lot of money repotting reject plants he brought home from work. We would look after them, revive them and eventually sell them back to smaller local nurseries. We did all the potting and watering and my father took care of the resale, splitting the profits straight down the middle with us.

After finishing university, I pursued a career as an actor, and did odd jobs in sales and marketing, but I missed working with my hands, so eventually I started working as a labourer with small private builders. I enjoyed the outdoor work again.

After a couple of years I was offered a job in the water-recycling industry, installing greywater systems in people's backyards across Melbourne, and rediscovered my passion for gardens. I soon moved into sales and promotion, and slowly found myself turning into what I now call an 'Aquavist' – a person who is passionate about water conservation.

Over the next few years, I learned the ins and outs of water in Australia. I found out about water recycling and rainwater-harvesting systems through trial and error. I talked to people in related industries at home and garden shows.

Eventually I did a lot of my own research and ended up travelling through parts of Europe and South-East Asia, discovering that we're not alone with our water problems in Australia. In Europe I learned that recycling was very common, even in areas with high rainfall, because the population is so dense. On top of that, countries like Denmark relied heavily on renewable energies, leaving me perplexed about our lack of progress in Australia. In South-East Asia, the issues were more disheartening.

Communities faced major water theft by large companies. Local people were powerless to stop it, and often saw no danger in signing away their water rights. Eventually I realised that the water shortage is a global issue affecting different countries in different ways.

I wanted to share this knowledge and do something positive for the environment. In early 2006, I decided to start my own company, helping ordinary Australian families and businesses take practical steps to save water. I started out small, relying on word of mouth.

At the same time, I began my Masters of Creative Writing at RMIT University. I wanted to write fiction, but I was encouraged by my lecturer, Antoni Jach, to write about the water issue instead. Antoni believes that you should only write about 'urgent and necessary' topics. He knew that I was passionate about the water crisis, so when it started to hit the front pages, he pushed me to write about it. After some initial research I realised there was no practical guide to saving water on bookshop shelves. Sure, there were brochures produced by environmental groups and water companies, but there was nothing definitive, no single book covering all the issues from saving water at home to climate change to rice farming.

Australians need this information. We want to understand the big issues, but most of us are looking for concrete, practical ways in which we can save water every day. We all want to be part of the solution to this problem – and that's why my co-writer Amy and I wrote this book. If every one of us could cut our annual water consumption by 50,000 litres, we'd have an extra 1000 gigalitres per year, or 400,000 Olympic swimming pools. We could use it to irrigate our farms, to safeguard the Murray, to preserve the future of our children. In this book, we'll show you how.

AMY'S STORY

I grew up in Canada. I lived in a rainforest where it poured like crazy. There was so much rain, the reservoirs overflowed. Then in June 2006 I moved to drought-stricken Australia to be with my fiancé. We lived together in a share house in Melbourne.

I had never timed my shower or considered recycling the contents of a stale glass of water on a thirsty plant. I could always get more out of the tap. I let the water run when I brushed my teeth and washed dishes one by one. This kind of behaviour didn't fly with my housemates. My partner had a pained look on his face when he told me that the water bill had spiked since I moved in. It didn't take me long to realise I had to change my ways and learn how to conserve.

I went to visit my in-laws at their home in Swan Hill, Victoria, and discovered that they go one step further than greywater recycling. They recycle their black water – the water from their toilet. The least I could do was start waxing my legs in order to cut down on my time in the shower.

I was a journalist in Canada, so curiosity is somewhat ingrained in me. When I met Craig in my creative writing class at RMIT, I was blown away by his wealth of knowledge about water. Water, the biggest issue in Australia. I had to know more. I was fascinated, and wanted to help him tell his story. It was my job to track down experts and conduct interviews. I helped out with the research, wrote some of the chapters and edited others. I learned a lot along the way – I hope you find the book as useful as I have.

INTRODUCTION

The vice-president of the World Bank famously said in 1995 that the wars of the twentieth century were fought over oil and that the wars of the twenty-first century will be fought over water.

We know we cannot live without water, but most of us can't imagine that we will ever reach the stage where we have to fight for it. Yet this is already a reality in some parts of the world – and not just in Third World countries. Globally, many people are already struggling to afford water for their families.

You only have to look at the US city of Detroit to see how precarious our access to water is. In 2001 Victor Mercado was named the new director of the Detroit Water and Sewage Department (DWSD). His background was with private water companies such as Thames Water and United Water, and many feared that his hardline business approach might lead to privatisation of the city's water. In fact it led to massive price hikes. Each residence was fitted with a water meter, and if customers failed to pay their bills, their water was cut off.

According to the Michigan Welfare Rights Organization, the DWSD cut water services to over 45,000 residences between June 2001 and January 2003. Residents were forced into the street. They had to fill bottles and buckets with water from public spaces. It was either that or rely on friends and family. The city's social services department removed children from more than 700 families because their parents could not afford to buy water to drink.

People gathered in the streets and camped out, not only in protest but in some cases out of necessity. Detroit became known as 'Tent City' and people begged the government to act.

The shut-offs affected low income earners the most and contributed to the increasing number of homeless people in Detroit.

Fortunately, the situation is looking up. The Michigan Welfare Rights Organization has had some success in pushing a 'Water Affordability Program' to the Detroit City Council, fighting to restore amenities to all citizens. After all, water is a basic human right.

All of this just seems ridiculous to Australians: it's like *Mad Max* or science fiction. It's hard for many of us to change our built-in belief that 'She'll be right, mate!' Australia has always been 'the lucky country' – but we now face a threat to the easy, prosperous life we're used to. Our weather patterns are changing. It rains less than it used to, and in different areas. Population growth is stretching our resources further than ever before. As pressure builds, we too will face price hikes. Water is liquid gold, and one day soon we may struggle to afford it – just like the people of Detroit.

In less than a dozen years, extreme measures will have to be taken to conserve Australia's water, and I'm not talking about mere restrictions. The government may be forced to ration water, allocating each of us a certain number of litres per week to live on. It may seem a bit unlikely now, but I assure you it could happen.

Australia faces serious temperature rises and significantly less rainfall by 2070, scientists are warning.

The Commonwealth Scientific and Industrial Research Organisation (CSIRO) has predicted that rainfall in parts of eastern Australia will drop 40 per cent over the next fifty or sixty years, and the average temperature will rise by 7 degrees. By 2030 the risk of bushfires will be higher, droughts more severe and rainfall and stream run-off lower. By 2070, the town of Gunnedah in western New South Wales will have more than 100 days a year with temperatures over 35 degrees, and Walgett, 300 kilometres to the north-east, may have more than eighty days a year above 40 degrees. Such constantly high temperatures could turn normally drought-proof green pastures into brown dustbowls.

Tough water restrictions are finally starting to make people realise how serious things are getting. Most states have now imposed the highest level of water restrictions in our history. South Australia, our driest state, has introduced permanent water conservation measures. The government is struggling to manage record low Murray River inflows after 2006 brought with it the driest winter since 1902. In Brisbane, Queensland, stage-five water restrictions have taken hold. That means no washing cars, watering cans only, no water pistols. Incredibly, I still hear people grumbling about not being able to water their lawns. I just don't understand.

In agriculture, sport, entertainment, business and the intellectual arena, Australians are a force to contend with. We compete at a world level and often win – and all from our big dry land at the bottom of the earth. So why are we so far behind when it comes to water conservation and recycling? European countries with much higher rainfall have been conserving water for up to thirty years. After Antarctica, Australia is the driest continent on the planet, but we're only just beginning to spark up about the water crisis we've already been facing for many years.

Part of our problem is that none of us want to take responsibility. City people often tell me that agriculture is to blame for the depletion of our waterways and reservoirs. People from country Australia complain that city dwellers are wasteful and ignorant – they've never had to survive on rainwater or groundwater, so as far as they are concerned, water is a substance that flows infinitely, and they simply find it at the end of the tap. But blaming each other is pointless. We are all part of the same cycle ... one large ecosystem. A country problem is a city problem and vice versa.

We all join together in criticising the government, of course. We wonder why our politicians aren't planning more dams, or investing in water recycling or desalination plants. The answer is that these are not simple solutions, and their financial and environmental costs are massive.

I think it's time we all stopped pointing the finger at others and took responsibility for our own water use. We can't wait around for someone else to fix the problem for us.

The aim of this book is to raise awareness of the water shortage and to explain how we can all do our bit in securing our future. There *are* solutions. Many of them are simple and practical, and they're not just cheap, they'll actually save you money. Best of all, they'll make you feel great, because you've done something for your family and the environment. I've put these tips right upfront, so you can get straight into them, but in later chapters I also look briefly at the politics of water provision and the various large-scale solutions governments are considering, so that anyone who wishes to get an idea of the bigger picture can.

It's not too late to fix our problems, as long as we act now. All we need is the right information. Public education really makes a difference. In Victoria, an initiative known as 'savewater!' has done an incredible job in reaching people right across the state. The campaign really took off a couple of years ago and since then water consumption in Victoria has decreased by about 100 litres per person per day. It's proof that people understand the issues and care enough to make changes to their daily lives.

This campaign and others like it give me hope for the future. I was seventeen when recycling systems were first introduced in Australia, and I remember quite clearly how long it took to for people to get used to separating cardboard and plastic and placing them in different coloured bins. At first no-one cared or could remember to do it, but now it's second nature to even the biggest lazybones.

I believe deeply in the power we all hold in our hands. Thousands of people around Australia today are employing a few basic ideas which have already saved millions of litres of water – you can too.

Remember: every last drop counts!

PART 1
WATER: THE FACTS

*It's not drought we're facing,
but climate change*

Australia is the driest populated continent on the earth. Only Antarctica receives less rain. When the storm clouds roll across Oz and let go, most of the precious rain that falls is lost to us. Tasmania, the Northern Territory and parts of north Queensland receive a large percentage of our total rainfall, but the bulk of our population lives elsewhere.

Drought has always been a fact of life in Australia. Older people still talk of the 'war years' drought which lasted from 1938 to 1945; until very recently, some even remembered the Federation drought. People living in rural communities are used to coping with a lack of water. We've faced drought before, so what's the difference now? It has to rain sometime, surely?

The reality is that we are no longer dealing with drought. 'Drought' is the word used to describe a *temporary* lack of rainfall. To say that we are now in our tenth year of drought is inaccurate. It's not drought we're facing, but climate change – a result of damage we have done to our planet, altering our environment almost beyond repair.

There is hope, though. We have faced situations like this before and found solutions. One of the best examples is the way the global community dealt with the problem of acid rain in the 1970s and 1980s.

When coal is burned to make electricity, it gives off sulfur dioxide particles. These particles mix with water vapour in the atmosphere to form sulphuric acid. Nitrogen emissions from cars and residential and commercial furnaces also mix with the water in the air and are transformed into ammonium nitrate and nitric acid. These substances travel in cloud form for hundreds or even thousands of kilometres, eventually falling back to earth as rain or snow.

Acid rain can 'kill' lakes, rivers and forests, along with the life forms they support. The phenomenon was first identified in the nineteenth century, but it was only when scientists began to make a serious study of the problem in the 1960s that people sat up and paid attention. As a result, international treaties aiming to reduce air pollution were negotiated; they came into force in the 1970s and 1980s. Emissions of sulphur and nitrogen have been reduced by more than 50 per cent since 1980, and acid rain is no longer the problem it once was.

Similarly, the international community came together in the late 1980s when it became clear that chlorofluorocarbons (CFCs) were ripping holes in the shield of ozone particles that circles the earth and absorbs harmful ultraviolet rays from the sun.

In 1987, worldwide efforts were made to reduce the use of CFCs within industry, and in 1996 they were banned completely. Scientists project that the ozone layer will gradually repair itself, and by 2050, the amount of CFCs in the atmosphere will have fallen to acceptable levels.

So, humans have managed to address global environmental problems in the past by changing our behaviour and acting cooperatively. I believe we can do it again – but first we need to understand the problem we're facing. It's difficult to just up and change merely because someone tells you to, no matter how much authority they have. My co-writer Amy wouldn't cut her glorious seven minutes in the shower down to four before she really understood why she should. Who would? If we're going to change the way we live, first we need to be convinced. We need simple, undeniable facts.

We are running out of water

Our water supplies in Australia are reaching record lows. The reservoirs feeding many of our towns and cities are at less than 40 per cent capacity – some of them much less.

Reservoir levels are falling because it isn't raining as regularly as it did in the past, or in the same areas. Weather patterns are changing and water does not seem to be making it to our catchments.

Let's take Sydney as an example. The total annual flow of water into all of Sydney's eleven dams averaged 71,635 megalitres between 1990 and 1996. (A megalitre is 1,000,000 litres.) By 2003 this annual average had dwindled by 45 per cent to 39,881 megalitres. As of mid 2005, Sydney's 4 million people had only two years' supply of water in storage.

This is a worrying figure, as Sydney's water storage capabilities are among the largest in the world – four times the size of New York's and nine times that of London, and still it seems it's not enough.

We're using more water than we ever have before. In the last twenty years our city populations have spiked up due to childbirth and immigration. We're having smaller families, but the population is still growing as a whole. People from the country are choosing to move to cities, too, where jobs and money are easier to come by.

We also use more water because our rate of consumption has increased. From the mid to late 1970s, our society's focus shifted toward material and monetary gain. The ideal became a big house with a swimming pool on an easy-to-manage block. It was a time to spend, to own and to consume. By the late eighties we'd forgotten the simple country life many of our ancestors led, and the frugal habits that helped them survive. Microwaves and dishwashers became the norm, and as technology advanced, we got fatter and greedier. By the mid nineties we no longer had any concept of conservation, or even moderation: all of the lights in the house stayed on at all times, we let the taps run while we brushed our teeth, and we became obsessed with cleanliness, filling our homes with chemical sprays and disinfectants. We're even worse today. Faster internet, bigger televisions, hotter water, bigger cars, bigger pools, more ... more ... more!

Burning fossil fuels leads to climate change

It's the buzz of electricity that makes our luxurious lives possible, but in generating it, we produce 'greenhouse gases' that are changing our climate and starving us of the water we need to feed our crops.

The greenhouse effect is a not a new phenomenon. Energy from the sun warms the earth, which then radiates some of the heat back into space. Naturally occurring gases in our atmosphere trap some of this heat and radiate it back to us. It is because of the existence of these greenhouse gases – including water vapour, carbon dioxide, methane and nitrous oxide – that the planet is as habitable as it is today. Without them the earth would be approximately 35 degrees cooler than it is.

The problem is that humans are tipping the natural balance, generating more greenhouse gases than the earth can handle. Since the industrial revolution, carbon dioxide emissions alone have increased by 30 per cent. The build-up of these gases means that more and more heat from the sun is trapped within our atmosphere, gradually heating up the earth's surface. This is commonly known as global warming.

Much of the electricity we use in Australia is created by burning coal, which produces carbon dioxide (CO_2). For every tonne of coal we burn, 1 to 2 tonnes of CO_2 are released into the atmosphere. Australians emit more greenhouse gas per capita than the people of any other country, averaging around 6.6 tonnes per person per year.

Another major contributor to global warming is logging. We have spent the last hundred years or so destroying forests to make room for agriculture and commercial and residential development. Australia has less than 10 per cent of its old growth forest left intact. We need our forests in order to take CO_2 out of the air for us and turn it into oxygen. The once mighty Amazon forest has actually become a carbon producing site.

Clear-felling reduces the amount of CO_2 a forest can absorb, while burning for land clearing releases more carbon into the atmosphere.

As earth's surface warms, so do the oceans. A very small increase in the temperature of the oceans will rapidly kill off small animal and plant life, such as plankton, threatening the entire food chain. What's more, marine plant life sucks up CO_2 and releases oxygen, just like land-based plant life. Nothing is more important to us and all animal life than oxygen.

As the oceans warm, the polar ice caps have started to melt faster than usual. According to NASA researchers, the North Pole has shrunk by 20 per cent since 1979.

In 2005, the first hybrid polar bear/brown bear cub was discovered in the wild in Canada. The caps are melting so much that the polar bears have been driven south and are mating with other species.

If the melting of the ice caps continues, the sea levels may start to rise, with potentially disastrous consequences.

Familiar weather patterns are already changing drastically. This is happening as I write. It is becoming obvious to scientists that global warming is causing the increasing numbers of tornados, hurricanes and tsunamis we have recently suffered and, at the other extreme, the extended periods of drought.

We can't keep building dams and praying that they will somehow fill. We must all do our bit by cutting our emissions and saving water. We can help by acting at the micro-level. Damage has been done, but it is not too late to undo it – yet.

What is being done about climate change?

The first major international conference on the greenhouse effect was held in Austria in 1985. Scientists spoke out about the magnitude of climate change the world was facing.

A few years later, in 1988, climatologists met in Toronto and called for a 20 per cent reduction of CO_2 emissions by 2005. The United Nations set up the Intergovernmental Panel on Climate Change.

At the Rio Earth Summit in 1992, 154 nations signed the UN Framework Convention on Climate Change (UNFCCC). Industrialised countries agreed that by 2000 they would cut their greenhouse gas emissions to achieve targets set below their 1990 emission levels. Australia signed and ratified the UNFCCC.

In 1997, an amendment was made to the UNFCCC. It was called the 'Kyoto Protocol', after the Japanese city in which the conference was held.

The protocol did two important things. First, it

assigned legally binding targets for emission cuts for developed countries, with a deadline of 2010. It also allowed for 'emissions trading' of the major greenhouse gases.

Before the Kyoto Protocol could come into effect, it had to be ratified by fifty-five signatory countries. Those fifty-five had to include industrialised countries accounting for slightly more than half of the total combined emissions of all the industrialised nations in 1990.

Australia signed the Kyoto Protocol in 1998, as did the United States, but neither country went on to ratify it. It finally came into force only in 2005, after Russia ratified it late in 2004.

The Howard Government still refuses to ratify the treaty, despite widespread public support for it, arguing that it would hurt Australia's economy (even though Australia would actually be permitted to increase greenhouse gas emissions by 8 per cent under the protocol). The government objects to the exemptions the protocol grants to developing countries, including China and India.

The United States has taken a similar line. The Clinton administration didn't submit the protocol to the US Senate for ratification because of concerns about its potential effect on the national economy, and because timetables and targets had not been set for developing nations. The Bush administration has followed suit, citing uncertainties about the true nature and extent of climate change as a further barrier to US participation.

There's a misconception that China is a massive polluter. It's true that China is the world's second biggest greenhouse gas producer and is expected to move ahead of the United States into the number one position soon, but it must be remembered that China's population is four times bigger than that of the US. China's per capita emission is actually quite low.

The initial reduction targets for industrialised countries were set at an average of only 5.2 per cent. China is aiming for a reduction of 15 per cent, even though as a developing nation it is not obliged to make reductions under the Kyoto Protocol.

In Australia, the Howard government has set our targets for reduction at 2 per cent of our 1990 emissions level. This is pitiful, considering that we are the planet's worst offenders, emitting 25 per cent more greenhouse gas per capita than the United States.

This is why I urge people to do their own bit. We can't wait for world governments to agree, we must reduce our own use and in turn our impact. The solution is simple: we need to cut down on the water and energy we use and switch to sustainable sources where we can.

We all have to take the facts on board and then apply this knowledge in our everyday lives. If 20 million people make small changes to their routine, it really adds up.

HOW MUCH WATER DO WE USE?

It takes 246 litres of water to produce a single glass of milk

We tend not to think about it, but the water we drink is a manufactured product – just like cornflakes or toothpaste or rubber tyres. It has to be harvested, processed, stored and distributed, just like any other commodity, and there are significant costs involved along the way. It actually takes 5 litres of water to produce a 1 litre bottle of water – and that's not including water used in transporting the bottled water from the factory to the customer.

We also forget how much water is used in manufacturing many other products we take for granted. For example, it takes 246 litres of water to produce a single glass of milk, because cows drink water, and it is also used to grow cattle feed.

Growing a watermelon would take around 378 litres of water. To produce an egg, you'd need 450 litres. A loaf of bread? 567 litres. A weekend newspaper? 570 litres. And a cotton T-shirt? About 4100 litres – that's around 12,000 glasses! Hmmm … it's enough to give anyone a dry throat.

Here are a few more shocking water facts and figures.

According to the CSIRO, it takes:
- 540–630 litres of water to produce 1 kilogram of maize
- 715–750 litres per kilogram of wheat grain
- 1550 litres per kilogram of paddy rice
- 1650–2200 litres per kilogram of soybeans
- 50,000–100,000 litres per kilogram of beef
- 170,000 litres per kilogram of clean wool

Source: www.clw.csiro.au/issues/water/water_for_food.html

Measuring water

In this book, I'll talk a lot about vast quantities of water – kilolitres, megalitres and gigalitres. Here's a quick guide to help you get your head around the huge volumes we'll be discussing.

1 kilolitre	=	1000 L
1 megalitre	=	1,000,000 L
1 gigalitre	=	1,000,000,000 L

Most of us have seen an Olympic swimming pool. They're 50 metres long, 25 metres wide, and a minimum of 2 metres deep. Assuming a depth of 2 metres, an Olympic swimming pool holds 2,500,000 litres – in other words, 2.5 megalitres.

Water trivia

Q. What are the three forms that water occurs in?
A. It occurs as a liquid, a solid and a gas – water, ice and steam.

Q. True or false: water is the only substance on earth that naturally occurs as a liquid, a solid and a gas.
A. True.

Q. How long can a person live without food?
A. More than a month.

Q. How long can a person live without water?
A. Approximately one week.

Q. How much water must a person consume per day to maintain good health?
A. A total of 2.8 litres from all sources, including water and food.

Q. How much water is used to flush a toilet?
A. Around 10 litres.

Q. How much water does the average shower head use per minute?

A. Between 15 and 25 litres.

Q. How much water does the average automatic dishwasher use on a full cycle?

A. 80 litres.

Q. How much water is used on average to wash a load of dishes by hand?

A. 20 litres.

Q. How much of the earth's surface is water?

A. 80 per cent.

Q. Of all the earth's water, how much is oceans or seas?

A. 97 per cent.

Q. How much of the world's fresh water is frozen and therefore unusable?

A. Two-thirds.

Q. How much of the earth's water is suitable for drinking?

A. Just 1 per cent. Yes, that's right – 3 per cent of the world's water is fresh and 2 per cent of that is frozen, leaving only 1 per cent available for use.

How much water do we use in our homes?

A little more than a half of all water supplied in our cities is used in and around the home. An average Aussie household uses about 260,000 litres of water per year. That's around 700 litres or four-and-a-half bathtubs full of water per day!

Here's the breakdown: kitchen, 10 per cent; laundry, 20 per cent; bathroom, 20 per cent; toilet, 20 per cent; garden, 30 per cent.

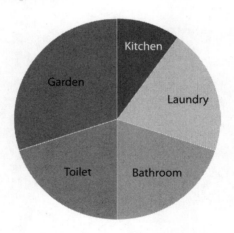

Water use in the average Australian home

In the next section of this book, we'll look at how you can save water both inside and outside your home. We'll start with some general tips, then go through each room, one by one.

PART 2
SAVING WATER AT HOME

AROUND THE HOUSE

electricity use = burning coal = greenhouse emissions = global warming = climate change = extreme drought

Before we look specifically at the kitchen, the bathroom, the toilet and the laundry, I'd like to take a general look at a few things we can do to save water at home.

It might seem strange to kick off a book about saving water with ideas about saving electricity, but that's what I'm going to do. By saving electricity, we save water – because we're helping slow down climate change. Then I'll share some waterwise tips that will help you save water in every room in the house.

Cutting down your energy use

It's easy to cut down on the energy you use. All you need to do is stop and think every now and again. We need to remember this equation:

electricity use = burning coal = greenhouse emissions = global warming = climate change = extreme drought

Calculate your carbon footprint

Before you try to cut the amount of energy you use, calculate your 'carbon footprint' – the amount of

CO_2 you generate every day. Google 'carbon footprint calculator' and see what you come up with. Online tools can help you figure out how much greenhouse gas you're putting out there now, so that you can look back in a few months and know just how much you're saving.

If you're really keen, you could have your home audited. Acting on an auditor's recommendations can help you reduce emissions by up to 50 per cent. You could also suggest a workplace audit.

Change to green power

Many power companies now offer green power options. They invest in alternatives to coal-fired power stations such as wind power and solar power, subsidising their research, development and construction costs by asking customers who want to limit their environmental impact to pay a premium. You can often choose to have a certain percentage of your power offset against a renewable supply. Depending on your budget, you might choose as little as 10 per cent or as much as 100 per cent.

GreenPower is a national accreditation scheme for renewable electricity products offered by energy suppliers around Australia. GreenPower's website says that since the scheme was first established in 1997, more than 380,000 domestic and commercial customers have saved over 3.4 million tonnes of greenhouse gas emissions – just by choosing to buy energy from accredited alternative sources. Visit http://greenpower.gov.au to find out how to make the switch.

If green power really appeals, you could take matters into your own hands, and get into domestic solar power. Most of us know that the sun's rays can be used to heat the water we use in our homes via north-facing panels placed on the roof. The systems are quite inexpensive and immediate results can be seen in your electricity bill. Many states in Australia actually give healthy rebates for the installation of a solar hot-water system.

Photovoltaic cells are another option. These are panels that convert the sun's rays into electricity, which must then be transformed using an inverter into an alternating current of the correct voltage for your area. These allow you to create your own power. You mount them on the roof just like a solar hot-water heater.

In some countries, like Japan, homeowners can make money by selling their excess solar power back to the grid. Up until now the cost of photovoltaic cells has been high, but it's falling. Within the next couple of years they will become extremely cost-effective.

Use energy wisely

The federal government announced early in 2007 that the old incandescent light globes are going to be phased out, but there's no need to wait until 2009 or 2010 – switch to energy-efficient light globes now! And while we're on the topic, there's no need to have a light on in a room when no-one is in it.

Energy-efficient whitegoods are definitely the way to go if you're replacing your old fridge or washing machine. Look at all the available models that meet your needs, then compare their energy-rating labels and choose the one that uses the least energy. The more stars an appliance has, the better. The numbers on energy-rating labels represent the amount of energy used in kilowatt hours per year.

For more information on the energy-rating labelling scheme, visit www.energyrating.gov.au. The website has information for all current models on the market, so you can compare them online before you go out shopping.

Another way to save energy is to make sure your house is properly insulated. Draw the curtains to keep the heat out during summer and in during winter, so you won't have to rely on air conditioning. If you're cold, put a coat or jumper on rather than switching on the heater, and use an old-fashioned door snake to block draughts.

The energy rating label: the more stars a product has, the more energy-efficient it is. The label also gives an estimate of how much energy an appliance will consume under standard conditions in kilowatt hours per year.

Transport

We should all be walking, cycling or catching public transport whenever and wherever we can. It's easy if you live in the inner city, of course, but if you're further out, you may need a car to get around. You can still do your bit, though. Try to think ahead, and plan your trips. If you need to drive to the shops, return your library books or videos at the same time. Do you live near a workmate? Think about carpooling to work. And if you're buying a new car, make fuel-efficiency your first priority. Hybrid cars are a great option these days, and they're getting cheaper.

Q&A *'Do I create more carbon emissions from my house or my car?' asks Nick, who has just moved from inner Sydney out to the suburbs.*

When calculating their 'carbon footprint', most people assume that the carbon they emit by burning petrol to power their vehicles is the major offender. It's always a good idea to leave the car at home and take public transport, of course, but the reality is that the electricity we use in the home generates far more greenhouse gas than our cars.

Write to your local member about climate change

When you know you've done everything you can to save electricity at home, don't stop there. Sit down and write a letter to your local MP letting her or him know that you care passionately about climate change. Public opinion drives public policy, so it's definitely worth telling your local representative what you think.

Being waterwise

There are a few simple things you can do at home that may help you save thousands of litres of water almost immediately. These are general-purpose tips that apply in the kitchen, bathroom, toilet, laundry and garden.

Deal with leaks promptly

A dripping tap may look like a slow, insignificant leak, but it can waste a couple of litres an hour. Let's say that's 48 litres a day. If you let a tap drip for a week, you'll lose 336 litres, and if it takes you a month to find the time to fix it, you'll have wasted 1488 litres of clean drinking water.

It's usually not hard to fix a tap, so why don't you have a go? All taps are different and they are fastened in a variety of ways. In some cases, an allen key is required to remove them. The good news is that once you get inside the taps, they're all basically the same. Here's what you do:

- First, take the head off the tap. You should see a spring around a brass spindle or shaft.
- Pull off the spring and then unscrew the brass spindle, which will leave you with a hole in the wall. Inside the hole in the wall is the washer. It will be made of plastic, and looks like a nipple. It has a rubber band on it.

- Use a pair of needle-nose pliers to extract the washer.
- Put a new one on, and then reverse the whole process. Easy!

Leaking pipes are another problem – and one that is harder to see. The best way to check whether you have leaky pipes is to turn off all taps in the house and read the water meter. Wait three hours (if you can go for three hours without using water) and read the meter again. If the numbers have moved, you have a leak. You'll have to find and fix it. If you're handy, you can repair it. If not, call a plumber. With any luck you know one who owes you a favour.

Get into the habit of checking your water meters on a weekly basis to identify any sudden changes in consumption that may indicate a leak. Repairing leaks will always save you money, even if you have to pay a plumber to solve the problem. If it's a hot water tap that's leaking, you're not just wasting water, but also gas or electricity.

Install aerators

Consider picking up an aerator from a local plumbing store. Aerators don't cost much, and they can reduce the flow of water from a tap by up to 50 per cent. They contain an o-ring (like a small washer) that remains in place when the flow from the tap is low but comes free and restricts the flow of water – without reducing pressure – when it is high. One company whose products I recommend is Water Wizz. Check them out at www.waterwizz.com.au.

Insulate your water pipes

If several people have a shower at your house every morning, you can waste a lot of water as you all stand waiting for the water to warm up. If you insulate your water pipes, the water in them stays warmer longer. It will also save electricity.

You should also make sure the thermostat on your hot-water system isn't set too high, so you won't have to add too much cold water.

Choose environmentally sustainable design

Some of us renovate only when we have to; others like to keep the kitchen and bathroom looking up-to-the-minute. Regardless of your approach, renovating is a chance to save water. If you're remodelling your bathroom, choose a smaller bath that's quicker and easier to fill, and hook it up to a greywater recycling system. New shower? Install a redirecting thermostat system that keeps the same water circulating in the pipes until it is hot enough to use. New toilet? Go for a dual-flush with a waterwise cistern – or maybe even a composting toilet. Opt for sustainable design whenever you have the opportunity. You'll save water and support companies that are doing their bit for our environment at the same time.

TOP TIPS FOR SAVING ELECTRICITY AND WATER AROUND THE HOUSE

- Always remember: electricity use = burning coal = greenhouse emissions = global warming = climate change = extreme drought.
- Calculate your carbon footprint, then cut down on your greenhouse gas emissions.
- Arrange an energy audit for your home or workplace.
- Choose accredited green power for your home.
- Install a solar hot-water system or solar panels.
- Generate your own electricity with photovoltaic cells.
- Switch to energy-efficient light globes.
- Use energy-efficient whitegoods. Check the energy-rating label and go for the model with the highest number of stars possible.
- Insulate your home.
- Put a coat on instead of using the electric heater.
- Walk, cycle or catch public transport whenever you can.

- Carpool to work.
- Check the fuel efficiency of a car before you buy it. If you can afford it, consider investing in a hybrid car.
- Monitor your water meter regularly.
- Fix dripping taps and leaky pipes immediately.
- Put aerators on your taps.
- Insulate your hot water pipes.
- Choose environmentally sustainable fixtures whenever you can.
- Write to your local member of parliament about climate change.

KITCHEN

*Kitchen taps run at an average of
20 litres per minute*

One day recently I was lazing in front of the Saturday TV cooking shows. Mister Jamie Oliver was whipping up something called a Botham burger (after Ian Botham, the English cricketer). I don't eat meat but I do love cricket, so I watched transfixed. Then, as he got the mince ready, he leaned over, turned the tap on full throttle and suggested to the viewer that it be left on throughout the process, as a little water on the hands helps to shape the minced patties. I completely lost interest in the cooking and spent the next two minutes squirming, staring at the unused flow. Finally I asked Jamie if he wouldn't mind turning the bloody tap off.

Of course he couldn't hear me!

Now why could he not have put just a cup of water in the bottom of a bowl? In the ninety seconds that he let the tap run, he wasted about thirty litres of water. Arrrgh!

Kitchens account for about 10 per cent of all water used in the home. If we make changes to the way we prepare food, wash our dishes and stock our pantry and refrigerator, we can drastically decrease the amount of water we use every day.

Shopping for food

If you want to save water in the kitchen, you have to start in the supermarket. We can save water – and energy – with our purchasing decisions. Every product has a story that unfolds before reaching us.

First, look at where a product is made. If it has been shipped a long way, considerable water and energy have been used to bring the product from the manufacturer to you, whether it's been transported by plane, boat, truck or train. The huge amounts of energy used to fuel these vehicles result in massive carbon output.

Next, you need to think about packaging. The more packaging a product has, the greater the amount of water and electricity used in its production. This all costs money and pushes up prices.

Finally, think about where you shop. If you have the option of buying from farmers' markets, you should. Not only is it great to buy food produced nearby, with a minimum of packaging, but you will also keep money in your local community.

Cooking

We waste a lot of water in the hurried business of preparing meals and feeding the family. The kitchen is a place where we should slow down and be conscious of what we do. When people say they are too rushed to think about how much water they use in the kitchen, I have to disagree. Being waterwise in the kitchen can actually save you time and improve the quality of your eating.

When washing or preparing food, just use a small amount of water in a bowl or pan – and then reuse it. Get creative. For example, if you're washing a leaf vegetable like spinach to remove the grit, use the same water to rinse peeled potatoes or carrots. When you've finished with the water ... out to the garden you may trot, and douse your saddest looking plant.

> **Craig's tip:** *When I'm making mashed potato, I put other vegetables in with the potatoes when I boil them, and strain them out before the potato breaks up.*

Q&A *'How can I save water when cooking?' asks Marie, a Perth pensioner.*

Well, Marie, it's very simple. You can start by using less water in the pot when boiling vegetables, pasta and rice. Vegetables don't need to be drowning in water when you cook them. Using less water saves electricity or gas, because the pot comes to the boil faster. When you're done, you can let the water sit until it's cool and feed the nutrient-rich liquid to your plants. Steaming vegetables is another great option: it requires very little water and it's much better for you.

Q&A *'I used to be a gourmet pizza chef in Richmond. One of our specialties was a prawn pizza. Every night the chef would make me defrost the prawns and fish under a running tap. The tap would run for about fifteen to twenty minutes and I never questioned it. What else could we have done?' asks Hamish.*

Prawns, scallops and squid don't usually need to be thawed unless you're going to prepare them in a special way before you cook them – for example, if you're going to stuff them or cover them in breadcrumbs.

If you do need to thaw them, the best way is to put them in a plastic bag and let them sit in a small bowl of water. About ten minutes is probably enough. The water should just cover them. So, Hamish, there's no need to let water run over the prawns – they'll defrost just as quickly by sitting there! This applies in a domestic kitchen as well, of course.

For further advice about handling frozen food, see www.foodscience.csiro.au/consumer.htm.

Commercial kitchens are notorious for wasting water, largely because they run under serious time restrictions. If you work in a kitchen, remind your bosses that it is in the interest of the company to reduce water use because it will save money. It's a matter of learning new habits. Once you get used to a new way of doing things, the kitchen will run as smoothly as before.

I've been a kitchenhand, so I know that washing down the kitchen at the end of the night also uses a lot of water. It's imperative for health reasons, not to mention a legal requirement. Still, cleaning up regularly during slow periods throughout the day will ensure that less water is wasted in the nightly wash-down. Taking care not to use excessive amounts of commercial detergents and floor cleaners also helps, because you won't need so much water to rinse it all off when you're done. Using too much detergent also increases the chance of someone slipping late at night in the rush to get home. Believe me, the waiting staff will appreciate it and so will the management.

Think before you turn on the tap

I'm sure we're all guilty of wrenching the taps on and overfilling a pot. I've caught myself filling the pot, then realising I've added too much and emptying some out, only to find I've dumped too much and have to add more water! We can save a lot of water just by stopping to think before we turn on the tap.

For example, when you're making tea or coffee, don't fill the kettle all the way to the top if you're only making one or two cups. This saves water, but it also reduces the amount of electricity you use, limiting your CO_2 emissions.

If you are running a tap, waiting for the water to get hot (or cold), catch the excess water in a bucket and use it on the garden. Kitchen taps run at an average of 20 litres per minute – you'll find you collect quite a bit. Keep a bottle of cold water in the fridge through summer so you don't have to run the taps until the water cools down every time you want a glass of water. And here's a hint for parents: don't let your kids drink out of a tap, as a lot of water is wasted this way. Encourage them always to drink from a glass. If they don't want it all, get them into the habit of throwing the rest on a pot plant.

Actually, there's a tip there for a lot of adults, too. Most people go to sleep with a glass of water beside their beds. They take a few sips before bed and maybe a couple upon waking. The half-full glass sits out until the next night, when they trot back to the kitchen, dump it down the drain, and pour a fresh one – and then the cycle is repeated. Why not tip the left over water onto the garden as you walk out the door in the morning? May not seem like a lot – it's probably only around 300 millilitres – but over a year, that adds up to over 100 litres of water.

It's also a good idea to install aerators on your kitchen taps. As I noted in the previous chapter, they reduce the flow of water by up to 50 per cent.

Washing the dishes

The best way to wash the dishes is to fill the sink and put the plug in. Don't just run the dishes under the tap! Only use as much water as you need – you don't have to fill the sink right up to the brim.

Half-fill the second sink with clean water and use that to rinse off the suds. That way you have control over the amount of water you use. If you only have one sink, place all the dishes on a rack and when you're done, pour a jug of water over them slowly to make sure all the detergent is washed away. We have a tendency to be heavy-handed with detergents. Using less detergent makes it easier to rinse off the suds – which means you won't need as much water.

Finally, make sure the plug in the sink fits properly, so you don't have to top up halfway between the cutlery and the saucepans.

Dishwashers

Dishwashers have improved dramatically over the last few years, and use far less water than they once did.

Older dishwashers use up to 80 litres of water on a full cycle, while their modern counterparts use as little as 15 litres per wash.

Q&A *'Do dishwashers use a lot more water than washing by hand?' asks Berryl, an interior designer from Sydney's south-east.*

In general they do, Berryl, though I have seen people leave the tap running while rinsing their dishes and that can add up fast.

Washing your dishes by hand is definitely the way to go if you can manage it – because you don't just save water, you also save electricity – but I know it's not always easy to find time.

The best of the new machines are almost as efficient as washing dishes by hand. Still, if you run the dishwasher when it's only half full, you're wasting water. Only use the dishwasher when you have a full load to do.

When buying a new dishwasher, check its water rating. Look for a WELS (Water Efficiency Labelling and Standards) rating sticker. These stickers help you compare the water efficiency of different brands and models. Every WELS product is given a star rating: products with six stars are the most efficient.

The WELS sticker also gives a water consumption

figure; for example, it might specify that a particular dishwasher uses 9 litres per wash. If a product claims to be water-efficient, but doesn't have a WELS rating, don't take the manufacturer's word for it. Always try to find out whether claims made about a product can be substantiated.

The WELS label: the more stars a product has, the more water efficient it is. The label also tells you how many litres an appliance uses per minute.

It's a good idea to check the energy rating too. (I explained how energy ratings work in the previous chapter.) Again, the more stars, the better.

For more information on these labelling schemes, go to www.waterrating.gov.au or check out www.energyrating.gov.au. Both sites list information for all models currently on the market, so you can compare them before you head out to make your purchase.

Garbage disposal units

The average garbage disposal unit uses about 30 litres of water per day. It doesn't seem like much, but it adds up. It could be up to 11,000 litres per year. These units also use electricity, of course, resulting in the production of greenhouse gas. They clog up the sewerage system, too, which translates to higher plumbing maintenance bills and puts pressure on treatment plants. Why not put all of your food scraps into a composting unit or worm farm to be used on the garden later? Makes sense!

Go online for more information:

- Find out how to make your own compost at www.environment.gov.au/education/publications/composting.html.

- Sustainable Gardening Australia explains the science of composting and does a good job of laying out the rules at www.sgaonline.org.au/info_science_of_composting.html.

- The Bokashi Bucket composting unit is ideal for

apartments and great for houses as well. Check it out at www.bokashi.com.au.

- The Permaculture Research Institute of Australia has a wealth of information on composting on its site: http://permaculture.org.au/?page_id=22.

> **Craig's dirty secret:** *One of my favourite water-saving tricks is to leave the mopping till a Sunday morning after the dishes have been done. The water left in the sink may be a little dirty, but it usually has a good amount of suds in it. I simply add a bit of methylated spirits and mop away. The floor always needs a dry mop afterward anyway, so the fact that the water is second-hand doesn't make much difference. Amy is horrified by this, and I understand that many readers will share her reaction, but the trick saves me about 20 litres of water! Every bit counts.*

Cleaning

There are several low-water and no-water cleaning products on the market now. They're available in supermarkets and online.

Microfibre cloths in particular can replace or greatly reduce the need for cleaning chemicals. They're made of minute fibres that lift dirt, dust and moisture from household surfaces. In my opinion, microfibre cloths do a better job than regular cloth. See our 'Recommended products' section at the end of the book for recommendations.

TOP TIPS FOR THE KITCHEN

- When you're shopping, buy local.

- Choose products with a minimum of packaging.

- Wash vegetables in a basin or tub – not under a running tap.

- Put just enough water in the pot to cover your vegies. Steaming vegetables in a small amount of water is even better.

- Don't defrost food with running water. Sit small items in a bowl of water.

- Think before you turn on the tap.

- Only boil as much water as you need when you make a cup of tea – you don't have to fill the kettle all the way up.

- If you're waiting for the water to get hot, catch the cold water in a bucket and use it on the garden.

- Keep a jug of cold water in the refrigerator so you don't waste water waiting for the tap water to cool.

- Don't drink from the tap – fill up a glass.

- Install an aerator on your kitchen taps.

- When you wash dishes by hand, half fill the second sink with enough water to rinse the suds off.

- If you have only one sink, put the dishes in a rack and rinse them by pouring a jug of water over the top.

- Use a small amount of detergent so that the dishes need less rinsing.

- Choose a water-efficient dishwasher with a high WELS rating.

- Only use the dishwasher when you have a full load.

- Avoid using the garbage disposal unit in your kitchen sink. Put food scraps in a composting unit or worm farm instead.

- Check out low-water and no-water cleaning products.

51,000 litres of water are used in the average Australian bathroom per year

For most people the bathroom serves two purposes: it is the room of groom and the room of relaxation. After a long day at work or looking after the kids, a cool splash revives. When we're sore, steam melds muscles into a more relaxed state. We can close our eyes and ... *ahhhhhh* ... wash our cares away. Sadly, this is a luxury we can no longer take for granted. Twenty per cent of the water we use in our homes is used in the bathroom. Ninety per cent of that is in the shower or bath, so by changing our shower habits we can save tens of thousands of litres per year.

How much water do you use in your shower?

Before I can give a household an accurate quote for a greywater recycling system, I have to figure out the amount of water they use in their shower. Even after all these years, I'm still amazed at how much water people waste in the bathroom. But I can tell you, I'm nowhere near as horrified as my customers are when I tell them how much water they use in a year.

I remember quoting a rather large system for one young family of four in a large home in Hillside, a very dry suburb in western Melbourne. They were in for a shock.

The kids bathed twice a day in a 500-litre spa bath; their father had a ten-minute shower morning and night; and their mother showered in two lots of fifteen minutes.

First I worked out their daily total, then a weekly total, and finally a yearly total. This is it:

2 baths at 500 L per bath	=	1000 L
2 ten-minute showers at 17 L per minute	=	340 L
2 fifteen-minute showers at 17 L per minute	=	510 L
Total daily use	=	1850 L
Total weekly use	=	12,950 L
Total yearly use	=	673,400 L

That's the equivalent of 2.7 Olympic swimming pools this family could potentially recycle each year. That would be enough water to keep their garden – and their next-door neighbours' gardens – alive all year round. OUCH!

You may be thinking there's no way *you* could possibly use that much water, but if you sit down and figure it out, you'll be surprised how much you waste – especially those of you with young families. Some of the worst offenders in the bathroom and shower are teenagers (especially boys). Any reader who has teenage children will certainly know what I'm talking about.

It's easy to work out how much water you use in the shower. The average shower head uses between 15 and 20 litres of water per minute. The modern AAA-rated shower heads vary, but on average they use about 9 litres per minute. Replacing your shower head alone could cut the amount of water you use in the bathroom by at least half, saving around about 20,000 litres of water per person per year. Yes, you heard me right: 20,000 litres!

This is why I consider the shower the main culprit when it comes to water use in an average home. (Although in the case of young families with top-loaders, it could be a fairly even battle between the two. See the 'Laundry' chapter for further details.)

Use the table below to find out how much water you use every time you have a shower. First, you'll need to figure out your shower head's flow rate. To check flow rates at home, all you need is a watch or timer and a standard household bucket. Hold the bucket to the shower head, turn the taps on at full pressure and let the bucket fill for ten seconds. If the bucket is one-third full, you've collected about 3 litres; if it's full, you've got about 9 litres. Multiply this figure by six and you'll have your flow rate per minute.

Litres per minute	Time in shower (minutes)											
	4	5	6	7	8	9	10	11	12	13	14	15
6	24	30	36	42	48	54	60	66	72	78	84	90
9	36	45	54	63	72	81	90	99	108	117	126	135
15	60	75	90	105	120	135	150	165	180	195	210	225
19	76	95	114	133	152	171	190	209	228	247	266	285

Once you know the flow rate, it's easy to calculate the amount you're using per shower: just multiply the flow rate by the number of minutes you spend in there. If you shower once a day, multiply your answer by 365, and you'll know how much water that adds up to per year.

Let's take Amy as an example. When I first met her, Amy used to have thirteen-minute showers, and she had a shower head that pumped out 15 litres per minute – that's 195 litres per shower. Multiply this by a minimum of seven showers a week, fifty-two weeks a year ... You do the maths! Yes, Amy was using over 70,000 litres of water in the shower per year. That is equivalent to filling one quarter of an Olympic swimming pool annually!

So, ladies and gentlemen of the long-shower brigade, how do you feel now?

Amy: I know, my shower habits were horrifying! I'm embarrassed. But I have changed my ways since I started working with Craig. I inspired my flatmates to install a AAA-rated shower head. It took some getting used to, but now I honestly don't even notice the difference. I can get in and out in just a couple of minutes if I'm not shaving or washing my hair. Otherwise, I still take about seven minutes – but I'm working on getting that down. It's not just a diet, I promise, it's a lifestyle change!

Having shorter showers can save you a lot of energy as well. The more hot water you use, the more gas or electricity you're consuming, so cutting down on your shower time means reduced greenhouse emissions and serious dollar savings.

The average Australian showers for eight minutes. Now, what I want to know is what the heck does someone actually do in the shower for that amount of time? It should only take four to five minutes to do what we need to do in the shower – maybe slightly longer for people with long, thick hair.

Until recently, most men have been good at keeping their showers short. A once-over with a bar of soap, a little bit of a stretch, a short ballad and they're out. I jump in the shower, get the job done and move on. For women, things can be a little bit different. On average, women tend to spend more time grooming and pruning than men do. They wash off make-up and shave their underarms, legs and bikini lines.

There are men who spend an unbelievable amount of time under the shower jet, though. Take my uncle, for example. He's a builder who showers for a good twenty minutes. There has been many a night when I have almost fallen asleep waiting for him to get out.

Men are becoming increasingly fashion conscious, too, and it's changing the way they behave in the bathroom. You spend longer in the shower if you are washing a half-kilo of product out of your hair every night!

In my opinion, it isn't necessary to wash hair more than once a week. Doing so will destroy the natural oils that keep hair healthy.

Amy: Sometimes Craig scares me. Please wash your hair more than once a week! Especially if you work out at the gym five days a week like I do.

I winced at the prospect of cutting back my thirteen-minute showers. I know I am not the only one who considers hydrotherapeutic me-time somewhat necessary.

But think about what all that hot water does to your skin and hair. It fries and dries. Be gentle on the earth and your looks and think about cutting your bathing time down. Stick to simply rinsing your hair some days followed by a rich moisturiser. Most hairdressers tell their clients they shouldn't shampoo daily. I find it does strip my hair of shine.

Q&A *'Which is better, a shower or a bath?'
asks Cara, a Brisbane barista.*

Fair question, Cara. It really depends on how long you shower for. A four- to five-minute shower definitely uses less water than a bath. If you refer to the table on page 62 you can see that even a high-flow shower head uses less than 100 litres if you limit your shower to five minutes. It takes about 150 to 200 litres to fill the average bath, so if you keep them short, showers are the way to go.

If you have children, though, the bath's a good option: you can use the same bathwater more than once. (Except when they're really filthy, of course.) Filling the bath uses up less water than two really short showers. But don't fill the bath all the way up – you don't need to! Once they're done, scoop the water into a bucket and put it on the garden or use it to flush the toilet. For very small children, use a baby bath.

> ## Talking water …
>
> I grind my teeth in frustration at the sight of an unused tap flowing away. The original cover of this book was supposed to be a faucet gushing water, expressing our carefree attitude toward this precious resource, but I could hardly bear to look at it.
>
> Not too long ago I was watching a repeat of *Queer Eye for the Straight Guy* (and no, I don't know why). One of the *Queer Eye* team was teaching a man how to apply hair product. Upon entering the bathroom, he immediately turned the tap on, and left it flowing at maximum knots while the scene unfolded. The segment went for about three minutes! Tick, tick, tick … 60 litres wasted, and for what?

This sort of behaviour is common in the bathroom. My old flatmate used to run the tap at three-quarter strength while she applied and removed her make-up! It absolutely enraged me, but I didn't know what to do. Finally I asked her if she realised that we were in a drought. She was insulted at first, until I explained to her about the dam levels in our part of Victoria and how rapidly they were falling, and eventually she saw my point.

It's really important to share your knowledge about the water issue, but you should try to discuss the problem with others – don't get up on your soapbox. (I know it's hard!) If you want to confront someone about their water use, try to turn the conversation around to the water crisis and find out what they know. Tell them facts about dams and rainfall in their own specific area, and let them persuade themselves. 'We're all in this together' is a much better approach than a lecture from on high.

Q&A *'Is it possible to get decent water pressure from a low-flow shower head?' asks John, a retiree living on the outskirts of Bendigo. 'We gave up on our water-saving shower head because it didn't feel like you were having a shower!'*

John is certainly not the only person I've heard this complaint from – but I have some good news: water-saving shower heads are much better than they used to be.

There's a difference between flow and pressure. Flow-rate is the *volume* of water that comes out of the shower head per minute. Pressure is the *force* of that water coming out. This means you can use less water without feeling as though you're showering under a single drip. You still get that nice warm-water massage while using less than half of the water you would with a non-restricted shower head.

A lot of people will be dubious, especially if they've had a bad experience with an older model, but it's time to try again. It's now possible to get decent pressure and save water at the same time. Please see the 'Recommended products' section at the back of this book for further details. Check out the 'Rebates' section at the same time: most Australian states offer rebates to people who buy a water-saving shower head.

> *'I don't want to use a water-saving shower head!'* says Alexandra, a musician in her twenties who frequently takes ten-minute showers. *'And I don't think I'm going to be able to convince the other people in my share house to invest in one.'*
>
> **Amy:** If you just can't stand the idea, try to cut the shower from ten to seven minutes. You can do it! And you'll have more time in the morning to decide what to wear. For about $10 you can buy cute little shower timers in different shapes and colours. After using one a few times you'll get into the habit. Think of the karma! You'll feel fantastic.

Q&A *'How much water do we waste while waiting for the shower to warm up? And what can we do about it?' asks George.*

Next time you have a shower, George, check the flow rate on your shower head and time how long it takes for the water to warm up. Then look at the table on page 62 and you'll have your answer. It will depend on the hot water available to you, whether or not someone has had a shower before you, and how far the water has to travel from the heating source to the shower rose.

I sympathise with George. Living in an apartment block with a shared hot water system as I do, it is frustrating to watch so much cold water washing away down the drain before I can jump under the shower. It takes a while to warm up because the water heater is a long way away from my bathroom.

Luckily, there are many things you can do to prevent this water from going to waste – and some are simple and inexpensive. For instance, you can put a bucket under the shower until the water warms up. Then use the water you've saved to flush the toilet, wash clothes or water the garden. You could also try to have everyone in the family shower one after the other, so the water stays warm.

These two tips are a great no-cost option for people who are renting or living in an apartment block, and can't undertake major plumbing projects, but homeowners can go further. Try insulating your water pipes, so that the water in them stays warmer longer. Another option is to install a 'redirecting' thermostat system that keeps the same water circulating in the pipes until it is hot enough to use. These systems are widely available these days, but when you buy one, be sure to find out what powers it. You don't want to save water and end up adding 10 per cent or more to your electricity bill – emitting more carbon gases while you're at it!

'I set our family's shower timer at three minutes each for a week and when we went back to five-minute showers it felt like an eternity, it was luxurious!' says George, a father of four in Melbourne's inner eastern suburbs.

Craig: If George can do it, the rest of us can at least try.

'People might think it sounds stupid, but for years I've been showering with a bucket. When I'm finished I take it out to the verandah and water the pot plants!' explains Ron from Nambour in south-east Queensland.

Amy: This is my favourite tip. So low maintenance! And there are lots of beautiful bath products on the market that won't hurt your blooms. See my reviews of the best biodegradable soaps, shampoos, conditioners and body scrubs at the end of this book.

Craig: If you're going to give this a go, make sure you use buckets big enough to stand in, as the water running down your body fills them up extremely quickly. I find a rectangular mop bucket the best. They allow you to capture at least four times more water than if you just sat the buckets in the shower to catch random drops.

'When I brush my teeth, I keep a glass of water by the sink and use that to rinse instead of letting the tap run,' explains Daniel, a Melbourne writer.

Craig: This is an oldie but a goodie. In all my years of running and attending water-saving seminars, this is usually one of the first points brought up. That just goes to show how many people do leave the tap running while they brush.

Be careful when you wash your hands, too. Don't just let water run. Half fill the basin instead. You could even keep a tub or container handy for this purpose, and then reuse the water you capture. Aerated taps are a good idea in the bathroom, just as they are in the kitchen, because they slow the flow.

> *'I think everyone should shower with a friend!'* says Richard from Ballarat, Victoria.

Craig: Yes, Richard, it's a great idea. But I'm not sure whether your focus is entirely on water saving. This solution mightn't be ideal for the singles out there, but for couples of consenting age, I have to say it's a fair proposition. Thank you, Richard!

Q&A *'I'm fairly indulgent when it comes to my showers, but I am disturbed by the amount of water I waste when I shave my legs,' says Alexandra. 'What can I do?'*

Amy: I started waxing recently and was amazed how much time it saved me. No more shaving! For those who don't like the idea of ripping hair from their body, just make sure you turn the taps off when you shave. Lather up first and then take your time with the razor. Rinse when you're done. You could even save the bucket of water you collected while the water warmed up at the start of your shower and use that to rinse your smooth legs! My draughty old terrace house is too cold for me to do this in winter, but it's easy enough in the warmer months.

We all need pampering and beauty time. I'm not suggesting you give all that up, but you can compromise. Lots of women don't shave at all. If that's too radical for you, why not consider this compromise: take a week off shaving every month in the winter. You will save time and water.

Men can do their part by growing a sexy beard and taking a break from the chore of hair removal. At the very least, guys, make sure you're not letting the tap run while you shave. Just fill the basin with hot water before you start.

> *'I hate those sensor taps in public toilets. They keep running after you've done washing your hands and you can't turn them off!' Lisa from Ballarat complains.*

Craig: The best thing to do with any public waste of water is to write a letter or email to the organisation responsible, whether it is private or public. Demand a response, and if you don't get one, make a complaint to your state's environmental protection authority. See the back of this book for contact numbers.

> *Dan, a Fitzroy footballer, says, 'The showers at the football club are turned on after a game or practice and they stay on until all the guys have rotated in and out of them and everyone is finished, instead of being turned off after each use!'*
>
> **Craig:** I suggest you be the one to have a word with the boys, Dan. I know from playing football and other team sports that people are exhausted at the end of the game; saving water is the last thing on their minds. You need to make them aware of the issue before they'll change their habits, but once you've convinced them it shouldn't take long. Maybe in a club with a fierce competitive spirit you could raise the stakes and give an award for environmental courage.

You don't have to be perfect and do everything all at once: just pick a few small things you can do to save water and make them a part of your daily routine. Make a change, feel good about it and tell your friends. Your actions have a ripple effect: if a few people you know follow your lead, and then their friends and family do the same, being waterwise will eventually become second nature for all of us, and before you know it, we won't have to worry about building new dams because we'll be living within our means.

Amy: Craig's right – it doesn't have to be all or nothing! If you really can't give up your ten-minute shower, just recognise it as a luxury and try to save the water in other areas instead. Suck it up and try a low-flow shower head, or take some responsibility for your lifestyle and invest in a greywater system. There are many tips in this book you can try.

Recycling the water from your shower or bath

Another good way to cut down on the amount of water you waste in the bathroom is to recycle it. If you install a greywater system, you can reuse the water from your shower in the garden. The average Australian garden uses even more water than the bathroom, so it's well worth considering. I've installed hundreds of these systems around Australia – they're becoming extremely popular, and don't cost as much as you'd expect. See the 'Recycling greywater at home' and 'Garden' chapters for full details.

If you are wondering why I haven't talked about the toilet yet, don't worry. I've devoted a whole chapter to the toilet, and it's coming up next. We use so much water in the toilet you could almost consider it another room!

TOP TIPS FOR THE BATHROOM

- Install a AAA-rated low-flow shower head.

- When you're waiting for the shower to warm up, catch the cool water in a bucket to reuse later. Better still, keep the bucket in the shower with you while you wash – you'll catch even more.

- Cut down your shower time.

- Buy a shower timer to help you keep your showers short.

- Wash your hair only when you need to – washing daily dries and fries!

- Take a short shower rather than a bath.

- Let your kids share a bath, or use the same bathwater and have them take turns.

- Have everyone in the family shower one after the other, so the water stays warm.

- Wax instead of shaving. Or use the 'warm-up' water from your shower to rinse afterwards.

- Only turn the tap on when you need to. Don't let it run while you shave or remove your make-up.

- Consider installing a redirecting thermostat system that keeps the same water circulating in the pipes until it is hot enough to use.

- Keep a glass of water by the sink and use that to rinse after brushing your teeth.

- Invest in a greywater system, and recycle the water from your bathroom in the garden.

- Report water wastage in public toilets to the responsible authority and don't rest till you've had a satisfactory response.

Twenty per cent of the water we use in our homes is flushed down the toilet

The Aussie outhouse has attracted the attention of jokers, storytellers and writers for over 200 years. Who hasn't looked under the toilet seat and checked for the famous red-back spider, and who has found one? Some of us have the great fortune to remember tiptoeing outside in the freezing cold in the wee hours and stumbling along a torch-lit path to answer nature's call. It's only in the last few decades or so that the toilet has become an integral part of the house and people began installing them indoors.

Once upon a time the old outhouse used no water at all and the sound of the dunny man's rattling buckets was a familiar one indeed. Night-cart men were cultural heroes who carried our dung on their shoulders, and they were paid very well for it. Their job has now been replaced by water! It's hard to believe, but 20 per cent of the water we use in our homes is flushed down the toilet. Fortunately, there are a whole range of ways to limit the amount of water you flush away.

Simple, low-cost suggestions

You don't have to spend much money – or any money at all – to save water in the toilet and take pressure off our already overloaded sewage treatment systems. Here are a few simple, low-cost suggestions.

Only flush the toilet when you really need to

When I visited my friend Bryce as a boy, his mother would always say, 'Don't forget to tell Craig the toilet rules.' Embarrassed, Bryce would take me into his room. 'What are the toilet rules?' I inquired the first time.

'If it's yellow, let it mellow ... if it's brown, flush it down.' Bryce's family was on tank water and understood how precious it really is. They depended on it.

Following that rule became second nature for me from then on. It was an affirmation of my destiny as an Aquavist.

It feels strange at first, not flushing every time, but you'll soon get used to it. Just keep the lid down and clean the bowl a bit more often. If you only flush the toilet when you really need to, you'll save up to 20,000 litres per year. Remember, *if it's yellow* ...

Some people might be uncomfortable leaving the toilet unflushed, but it can make a huge difference. Even if you only flush every second time you urinate, you'll save 50 per cent of the water wasted in the toilet.

For example, if you ordinarily flush the toilet six times a day, you could save 30 litres per day or up to 11,000 litres of water per year. Depending on the type of toilet you have, you can save up to 20,000 litres of water a year by only flushing once a day!

Use the warm-up water from your shower to flush the toilet
If you've read the 'Bathroom' chapter already, and you're collecting the 'warm-up water' from your shower in a bucket, use it to flush the toilet.

Modify your old-style toilet to use less water
If you have an old-style single-flush toilet with a huge cistern, buy a touch-sensitive tube to insert in the cistern under the button. They let you control how much water you use: the longer you hold down the button, the longer the flush. You can buy these devices from local hardware stores or plumbing stores. They're manufactured by Nylex and cost about $15.

People sometimes ask me about the old trick of putting a brick in the cistern so it takes less water to fill. This will work, though cisterns are very sensitive in their workings and any little knock or small piece of debris from the brick may cause damage and stop your toilet from flushing or filling properly. An old plastic juice bottle full of water is probably a better alternative.

Check your cistern for leaks

A leaking toilet can waste up to 200 litres per day. Unfortunately, it's very difficult to tell you have a problem when the leak is only small, because the trickle of water isn't clearly visible against the white porcelain of the bowl. Check whether your toilet is leaking by placing some food dye in the cistern and wait to see if it leeches into the bowl. If you detect a leak, call a plumber to fix it straight away. The unseen leak is the most wasteful, so I consider this one of the most important tips!

Choose eco-friendly toilet paper, pads and tampons

When you buy toilet paper, you can use your dollar to save water and support companies that do the same. The production of 'recycled' toilet paper requires 55 per cent less water than regular toilet paper. Recycled products are also cheaper because of their lower production costs. Locally made products are the best, as they don't have to be transported as far and so have less environmental impact. Always choose unbleached paper if you can, as the bleaching process pollutes our waterways with dioxins – chemicals that are poisonous to plant and aquatic life.

Merino Pty Ltd was the first company in Australia to manufacture environmentally responsible paper products, beginning with unbleached toilet tissue, sold under the SAFE brand. There are a number of other recycled toilet tissue brands available now – see the 'Recommended products' section for details. Generic brands such as Black & Gold, Payless and No Frills are often made from 100 per cent recycled paper, but they may not be unbleached – you should always check.

Women might consider trying tampons or pads produced from organic cotton without the use of fungicides, herbicides, pesticides or bleaches. They're available at selected health stores and pharmacies in Australia, and from a huge number of online stores. Never put pads or tampons down the toilet, as they create problems for our already overloaded treatment systems. They belong in the bin.

Don't turn the light on in the toilet unless you need to

There's often no need to turn the light on in the toilet. This might seem like an odd tip, but remember: using electricity generates CO_2, contributing to climate change.

Alternative toilet technologies

The water-saving solutions we've discussed so far in this chapter are all simple and fairly cheap, but if you own your own home, you can go a bit further and make some bigger changes. Until recently, there weren't many mainstream technologies aimed at saving toilet water. You had to actively search for such products or be lucky enough to come across them. These days, you have heaps of options, including dual-flush toilets, greywater and rain-flush systems and composting toilets.

Dual-flush toilets

Dual-flush toilets have been around for some time. They let you choose between a full flush or a half-flush. Unfortunately, older model half-flush toilets still use a fair amount of water. An older single-flush toilet uses up to 15 litres; the early dual-flush toilets brought this down to about 7 litres on a half-flush.

Q&A *Ian, from Carlton, Victoria, asks, 'Are there moves afoot to reduce the size of cisterns?'*

Good question, Ian. In 2004, Caroma set the standard when they introduced a dual-flush toilet that uses only 4.5 litres per full flush and 3 litres per half-flush. These systems can save the average person up to about 18,000 litres of water per year. That's a lot of water! A family of four would save up to 72,000 litres and $720 a year. The savings on water alone would pay for a new toilet in almost no time at all. I haven't found any other toilet manufacturers who are doing as well as

Caroma in building water-saving, dual-flush models, but new units are coming out all the time.

Greywater recycling

It doesn't cost much to install a greywater system to recycle wastewater from your bathroom and use it to flush your toilet. I've devoted a whole chapter to the subject. Please see the 'Recycling greywater at home' chapter for more information.

Rain-flush systems

Rain-flush systems are a great water-saving option. A pump connects your toilet cistern to both the existing mains line and your rainwater tank, so that the cistern can be refilled with rainwater from the roof when it's available. The real beauty of the system is that it senses when the rainwater tank is empty and automatically switches back to a mains supply – there's no need for manual interference. Rain-flush systems can also be hooked up to the washing machine. A rain-flush system could help save a household up to 70,000 litres of water per year.

Several manufacturers produce these systems. One I recommend is Davey, an Australian-owned company.

See the 'Rainwater harvesting' chapter for more details about rain-flush systems. Check the 'Rebate' section at the back of the book, too, to find out if you might be able to claim some of your purchase or installation costs back.

Composting toilets

Composting toilets aren't exactly a new invention, but they are only just starting to become popular in Australia.

> Don, a Canadian-born environmentalist and writer, has his say: 'Using something as precious as water to get rid of waste is ridiculous. We should make like the Swedish and use self-composting toilets!'

There are many different models of composting toilets, but they all work on the same principle as an ordinary garden compost bin. Waste is collected in the composting chamber, which contains carbon-rich material such as wood shavings and garden clippings. The organic materials gradually decompose in the chamber's aerated environment. The composting process is entirely natural.

The broken down waste can then be used as

nutrient-rich compost for your garden, if sufficient time is allowed and correct treatment conditions have been maintained.

Only buy a model that meets the Australian Standard for Composting Toilets.

A great benefit of the composting toilet is that you can throw any organic matter into the toilet and it will compost with the rest of the waste. Experts recommend that you throw some paper in as well, to improve the consistency of the compost. So from teabags to orange peel, having a composting toilet can be a bit like having your own recycling plant.

Best of all, a composting toilet can save a family of four up to 150,000 litres of water per year! Yes, that's the right number of zeros. That equates to savings of around $1500 annually.

People often ask me if composting toilets smell. The answer is no: an exhaust is set up to ensure that absolutely no smell escapes into the house.

The next thing they ask is 'Is it safe to use human waste in the garden?' Putting raw sewage on the garden would be dangerous, of course, but the compost from your toilet is not raw sewage. The harmful bacteria in the faecal matter is broken down by other, good bacteria, and the result is no different from any other nutrient-rich compost.

> *Ronald from Queensland has used a composting toilet for eight years, and he couldn't be happier. 'We looked around at the options for water saving before we built our house. We're not on town water, so it was particularly important for us. The composting toilet was a little expensive at the time but I believe it has paid itself off by now. They're a lot cheaper these days, too, because of higher demand and market competition. We are so happy with the way it works and so is the garden, which benefits from the compost that comes out. The fruit trees absolutely love it and it has allowed me to have a thriving garden even in drought conditions. I've been a passionate gardener my whole life and I can honestly say that this is the healthiest garden I've ever had.'*

Before having a composting toilet installed you should do your research, as they don't suit everyone. They are great on two conditions:

- You will need a house on stumps so there is room to dig a hole for the waste to settle into.
- You must have somewhere to put the nutrient-rich compost the toilet produces – a garden.

If you want more information on waterless and composting toilets, go to www.greenhouse.gov.au/yourhome/technical/fs27.htm.

Commercial dunnies

For years I have scowled at people in the MCG toilets during games. Not because their team was beating mine (when you support Richmond that's a given), but because at least 50 per cent of the taps are left running after people have finished washing their hands. Fair enough, there's a steady flow of people at intervals, and everyone's in a rush to get to the bar and then back to the game, but how much time does it take to turn off a tap?

Imagine how many people use the toilets in the breaks, how many taps are left running, and how many litres of water are wasted over a three-hour period! It would be in the range of 10,000 to 20,000 litres of water.

For me, the solution is easy: impose a fine on anyone who leaves a tap running.

The management in public venues could put up signs. We have no-smoking signs everywhere, so why can't we have signs that threaten fines for not turning off the taps? Enforcing the fines would soon sort the situation out. Not only would an organisation save on water rates, but it'd also be bringing in revenue.

Constantly leaking urinals are another issue – and it's not only football clubs that have those. Solutions to this problem are out there: a newly emerging technology is the waterless urinal.

Waterless urinals work on a very simple principle. Underneath the urinal is a layer of vegetable oil that acts as an odour trap, so that the smell cannot escape. The urine seeps through the trap and into a sewer pipe that feeds into the main sewerage system.

The Sydney Convention and Exhibition Centre has installed sixty of these new urinals. The original urinal troughs at the centre were flushed automatically with sensors, using between 8 and 15 litres of water per flush, and many of them leaked.

Mike Byrne, the director of Watersave Australia, says that the convention centre will save up to 4.5 million litres (that's eighteen Olympic swimming pools) of water per year, just by installing the waterless urinals. Brilliant! This is a great example of how industry and business can take the lead in tackling our water problem. Things can and are being done on a commercial level to limit the amount of water flushed down our toilets.

Composting toilets are heaven at music festivals

Amy: I went to the 2006 Meredith Music Festival in Victoria prepared to brave a weekend of stinky, wet and grungy port-a-loos. When I arrived, I was amazed to see a bank of odourless, waterless toilets. Each stall had a bucket of woodchips and a sign asking people to throw a scoop down the toilet when they were done. All I could smell was wood!

The waste was eventually used by Meredith area nurseries.

Composting toilets are the best thing that could ever happen to a festival! They're supplied by Natural Event, an Aussie company. The Natural Event toilets save 55 per cent of water over conventional toilets and produce compostable solids for the garden. They've been used at the Nati Frinj festival in Natimuk and the Falls Festival in Lorne, both in Victoria, and at Marion Bay in Tasmania. Natural Event is also getting into the trance/electronic circuit with the Rainbow Serpent and Akasha festivals.

TOP TIPS FOR THE TOILET

- Only flush the toilet when you really need to. Remember, if it's yellow …

- Use the 'warm-up water' from your shower to flush the toilet.

- If you have an old-style toilet, buy a touch-sensitive tube to insert in the cistern under the button so you can control the volume of water when you flush.

- Check whether your toilet is leaking by placing some food dye in the cistern. If it's leaking, get it fixed!

- Use unbleached, 'recycled' toilet paper, like SAFE.

- Only turn the light on in the toilet when you need to.

- Consider installing a dual-flush toilet.

- Use greywater from your bathroom to flush your toilet.

- Investigate a rain-flush system – you can use rain water from the roof or the water from your washing machine to flush your toilet.

- Invest in a composting toilet. Turn your waste into compost for your garden!
- If you run a small business, or if you're involved in managing public facilities, do some research into water-saving technologies.

An average top-loading washing machine uses 200 litres of water per wash!

These days it's hard to shock me – I've seen people wasting obscene amounts of water in their homes – but I was still horrified when I spoke to Betty, who was renting an apartment in St Kilda. Betty told me that she had just done a single-item load in the washing machines in her apartment block's communal laundry.

'What kind of item did you wash by itself?' I asked.

'I am embarrassed to say, it was a tea towel!' Betty winced.

Then she confessed that it wasn't the first time.

It turned out that she was doing this at least once a week. Now I winced!

I asked her why she was doing this regularly.

'Well,' she said, 'I just hate having anything dirty. I just can't wait until it all piles up! I live in a rented apartment and I don't have much space ... I just never thought about it.'

This sort of behaviour really is a disgrace, but before I spout off, I must admit that I have been guilty of washing a load with only one item in the machine. It was many years ago, before I had become aware of Australia's growing water issues. I could never do it now. Once you understand how serious our situation is, you just can't waste water like that and feel good about yourself.

Until recently, apartment-dwellers haven't paid directly for the water they use. It's only when separate water meters are connected for each flat that tenants can be billed for their water use. With shared water services in a lot of older apartment blocks, the cost is built into the rental price instead. It's understandable that people who have never seen a water bill and rarely escape the city might be ignorant of the cost of water and the effect of drought, but just because you don't own the place you live in doesn't mean that you can avoid responsibility.

The water shortage affects all of us and I daresay that when prices go up it will be the people who can't afford a house of their own who will suffer the most.

What could Betty do instead of washing items one by one?

She could place a basket in her closet and wait until she has filled it with dirty clothes before doing a load. She could also wash individual articles by hand in a half-bucket of water and rinse them in another. Washing a tea towel in a bucket uses only around 9 litres, rather than 160.

In fact, after much berating from me, Betty has started to make some changes. She makes sure she has at least a half-load before she does a wash, and she now adjusts the water setting on the communal washing machine to 'low' if she's doing a small load. It takes time to change our habits, but it's important that we start.

I still tease Betty about her single-item loads. Not many people are as irresponsible as she once was, but a lot of households use more water in the laundry than any other room in the house. It's not uncommon for a family with young children to do more than ten loads of washing per week. I've met families who do fifteen or twenty loads a week in a top-loading machine. If you have more than three children you'll know why!

Fifteen loads of washing in a top-loader uses approximately 3000 litres per week. Imagine 333 buckets of water. This would be enough to keep your garden and all your neighbours' gardens alive in summer!

Saving water in the laundry

On average the laundry is responsible for 15 per cent of our domestic water use. This makes it an ideal place to save water, and there are lots of ways to go about it.

Choose a front-loading washing machine

Installing a front-loader is one of the best ways to save water. Water-efficient front-loading machines are common these days. They often use a lot less electricity than top-loaders and 50 to 70 per cent less water, too. The problem is that front-loaders are generally not as quick as the good old water-guzzling Kleenmaid or Westinghouse models, which only take around 45 minutes per load. Front-loaders can take up to two hours on a full cycle. They're also more expensive to buy than top-loaders, but it's definitely worth switching. A front-loader will save you money on your water and electricity bills in the long run.

Q&A *Charisse asks, 'How much does it cost to buy a water-efficient front-loader?'*

A washing machine's cost depends on its quality, as well as its water and energy efficiency. The better its pump is, the longer the machine will last.

Water-efficient machines range from around $800 to $2000. The average mid-range machine will cost you about $1000. The average six-kilo front-loader uses between 50 and 80 litres for a full wash – roughly three times less than the average top-loader. Some smaller front loaders use as little as 30 litres per wash.

Like dishwashers, washing machines sold in Australia are given water-efficiency and energy efficiency rankings. It's best to buy the machine with the highest rating possible. (See page 50 in the 'Kitchen' chapter for further details about these ranking schemes.)

When buying a machine don't just rely on the manufacturer's claims – check what they are based on. There are fines for giving misleading information about water and energy efficiencies, so it is in the best interest of the manufacturer to do the right thing, but an independent rating like WELS allows you to compare different models much more easily.

I know that most people only buy new whitegoods when they really have to and that I can understand. Until your top-loader gives up the ghost, you'll be reluctant to spend money on a new one. In the meantime, you can save water in other ways – by handwashing rather than doing small loads, for example, or recycling your used laundry water on the garden. The same goes for people who rent. When you share a washing machine with others, the landlord is usually the one who decides which make and model to install. Whatever your situation, there are still some basic things you can do to reduce water waste. Number one: Don't adopt Betty's approach to laundry!

Greywater harvesting

Washing machines pump used water out quickly, allowing us to spread it around the garden without a lot of new technology or expense required. It's easy to set your machine up to drain into a plugged sink and then use a bucket to carry the water out to the garden. Just make sure you use a garden-safe detergent that is low in phosphates.

If you want to go a step further, you can invest in a greywater diversion system. The one real advantage top-loading washing machines have over front-loaders is that they make it easier to recycle your greywater on to your garden cheaply. The greater the volume of water you're using, the larger the area you are able to water. I'll go further into the details in the 'Recycling greywater at home' chapter.

Handwashing

If you've only got a few items to wash, it may well be better to wash them by hand. Handwashing uses no electricity, saving you money and reducing your impact on the environment.

You'll only need a few buckets of water. The big problem is how much you use for the rinse when you're done. Running the tap over the clothes is not the answer. For a few items, you should need no more than two fresh buckets of water to rinse out the suds. If you need more, you've used too much detergent in the first place.

If the stains are hard to remove, soak the clothes in NapiSan overnight to loosen and lift the dirt, then just rinse as normal. There's no need to overdo the NapiSan.

It's the amount of time you allow the item to soak that matters.

Please note that handwashing for its own sake is not necessarily more efficient. If you only have a few items and you need to do them straight away, it's the best option, but if you can wait, it's probably more efficient to do a full load – as long as you have a front-loading machine. If you have a top-loader, that's different. Handwashing is always more water-efficient than using a top-loader – and you don't use any electricity, either!

Use low-impact soaps and detergents

Once you've done a wash, the water goes direct from the washing machine hose to the sewerage system and eventually winds up in one of two places: either in your own septic tank (if you live outside the city) or at a large treatment plant.

A majority of Australians live in our cities – around 65 per cent – so our cities' treatment plants are under a lot of pressure. Our wastewater must be treated before it can safely be used for irrigation on farms or pumped into the ocean. The more chemicals you use at home, the more difficult it is to treat the wastewater at the other end, and that requires more money and electricity. As we already know, more electricity means more greenhouse gas ... and the global warming cycle continues. So, when you use chemicals, try to think about the effect they might have outside your home.

Bleach products containing chlorine and sodium can do real damage to our environment, and they should definitely be avoided if you plan to recycle your laundry water onto your garden. Some detergents also have high levels of phosphates; again, these can harm your plants. See the 'Recycling greywater at home' chapter for more detail, then check out the 'Recommended products' section at the back of the book.

Q&A *Caroline from Melbourne presents a tough question: 'What's better for the environment and water conservation – disposable nappies or cloth nappies?'*

There are arguments for and against both. Cloth nappies are made mostly from cotton, a product that takes a lot of water to produce, and people tend to use a lot of bleach and other chemicals when they wash them. Disposables are much easier to use, but they put pressure on landfills.

I'd say that cloth nappies are better in the long run, if you handwash them, but not many people want to do that – even if they have the time. A nappy wash service is a good compromise. There are lots of these services out there and they cost about the same as disposable diapers. It's worth calling the laundry businesses in your area to see who is making an effort to save water. Princes Laundry, based in Braeside in Melbourne's south-east, has installed water-recycling technology that has cut its water use by 40 per cent. Its nappy wash service delivers to 1000 homes a week.

Reusing greywater from nappy washing is not recommended, but laundries can make savings by washing nappies in bulk, in recycled water that has been used to clean clothing.

Q&A *'My sons play cricket, so I use a lot of chemicals to wash their clothes that I know are bad for the environment. Is there anything else I could use that is biodegradable?' asks Dianne from Mount Gambier in South Australia.*

I'm not an expert on stains, but I do play cricket, so this one I know. If you get to them quick enough, some vinegar, bicarb soda and maybe a little lemon juice can fix those grass stains.

If you do need to use some NapiSan, make a paste and use as little as possible. Apply it to the stain for fifteen minutes or so before soaking. In this sparkling, keep-up-with-the-Joneses world we feel our whites must be whiter than white, but ask yourself how much is enough and use chemicals in moderation. A whole bottle of bleach won't clean an item more than a tablespoon.

You old dog!

If you wash your dog in the laundry, just put a bit of soapy water in the trough. Don't let the water run over the pup for ages while you're scrubbing away. Go outside for the rinse and use a trigger nozzle on the end of your hose to get all the suds off. An outside wash is essential for big dogs anyway.

TOP TIPS FOR YOUR LAUNDRY

- Only wash when you have a full load to do. It saves electricity as well as water – and dollars and carbon emissions too.

- Use the 'economy' setting on your washing machine.

- Adjust the water level to suit the size of the load.

- When you're buying a new washing machine, look for the WELS rating and choose the most water-efficient model that meets your needs.

- A front-loader is better than a top-loader. They use from 50 to 70 per cent less water.

- If you only want to wash a couple of items, and can't wait till you have a full load, handwash them in a bucket.

- Set your machine up to drain into a plugged sink and then use a bucket to carry the water out to the garden.

- Install a greywater diversion system to pump your used laundry water onto the garden.

- Use a garden-safe detergent that is low in phosphates.

- Limit your use of bleaches and other chemicals.

- Wash small dogs with a little soapy water in the laundry trough, then go outside to the garden and use a trigger hose to rinse them off.

The average garden hose pumps out around 1000 litres per hour!

I was recently at a friend's birthday party in Queensland and got talking about water restrictions. (People are always having to put up with my water talk.) My ears pricked when I heard my friend's brother say: 'I have a very cheap water-saving idea for the garden.'

'Well, you know those little stickers they give to households for using recycled water on your garden, all you need is one of those.' He laughed, I squirmed. 'I mean who is going to know? Really, who is going to check?'

He was talking about the signs you see posted in parks, gardens and yards that explain that a rain tank or greywater system is being used. If you run recycled water through a sprinkler at all hours of the day and have a blooming garden in a time of drought, you really need one of these signs. Otherwise suspicious neighbours might think you're ignoring water restrictions and dob you in.

The manufacturer usually gives you a sign or sticker when you buy a greywater system. You can also get them from the person who installs a rain tank or greywater system for you. They're stocked at major plumbing and hardware stores, too. These water-recycling signs are like a badge of honour for responsible citizens. The thought of using one as a license to waste H_2O made me see red.

Now before jumping the table to strangle him, I tried to figure out whether he was joking or not.

He wasn't.

I took a deep breath and faced the fact that many people don't understand the dire situation we face.

⬥

We are an outdoors nation. Most Australians enjoy having a garden. Some grow fruit trees for produce and vegies for the family. Others dig in the dirt, coddling blooms for pure delight or peace of mind. Our gardens are important for our survival, too. We need plants to photosynthesise and create oxygen for us to breathe. If we all cemented our yards in that oh-so-beautiful 1970s way, we'd be all dry in the throat. So, in order to keep our gardens alive, we need to be more water-efficient. Our drinking water is too precious to waste.

The garden is the biggest water guzzler at home. In fact, one in three litres of our drinking water goes to feeding our gardens. Even if we were to focus only on the garden in our water-saving measures, we could combat the crisis effectively and secure the future of our drinking water supplies. By creating a sustainable, waterwise garden, a household can save between 100,000 and 200,000 litres of water per year.

People often ask me 'How do I keep my garden alive while cutting back on my mains water use?' I love it when I'm asked this question. There are many solutions available to the average Australian household. Choose from the simple plans in this chapter and cut that whopping 30 per cent average down to less than 10 per cent – or maybe even down to zero!

Caring for your soil

Healthy soil is the key to a healthy garden. Here are a few tips to help you keep your soil nice and moist, without using too much water.

Wetting agents

Wetting agents help soil retain water for long periods of time, so that plants can suck it up on demand, and you don't have to water so often. It is easier for water to reach its target – a plant's roots – when the soil is already wet. Moisture in the soil travels through tiny channels or 'watercourses' created by water that has flowed through before, just like the water from rain clouds that flows down mountain tops into rivers and out to sea.

After a long dry period, watercourses may need to be reformed, and it may take a few good soakings before the watercourse is reopened or another course is formed. This is why you may hear people say that the best time to water is after a rain.

Wetting agents are designed to keep existing watercourses intact during dry periods, so that the next time you give your plants a drink, the liquid will be sure to reach their roots. If you reapply a wetting agent every six months, the waterways in your soil should remain open.

There are a few different wetting agents on the market. Water crystals or granular wetting agents should be placed around a plant's root system, preferably at the time of planting. The crystals actually expand as they soak up water and then store it, creating a mini reserve the plant can draw on as it needs to. The benefit of these crystals is that you can give an area a good water and then forget about it for many weeks. There are several different brands and forms of water crystals available. Wetting agents are available at most garden supply stores and nurseries. Well-known names include SaturAid, MoisturAid, and Hydraflo2.

Apply wetting agents before summer – sprinkle the product on the surface of your garden beds or mix it in when planting. Reapply every six months or more regularly if required. You can also use granular wetting agents in containers, pots and baskets or over the entire garden and lawn, paying special attention to dry spots.

Compost is a natural wetting agent. The organic matter breaks down into the soil, keeping it healthy and moist, allowing the watercourse to stay open. Let the compost break down a bit first, then work it into your soil, where it can do some good.

Mulch

Mulching is one of the most effective ways to reduce water use in the garden. The average household uses around 1600 litres of water in the garden every week, but most of it evaporates before it can do any good. Efficient mulching can reduce evaporation by up to 70 per cent, and help you cut that weekly average of 1600 litres down to just 480 litres per week.

Mulching is good for your plants, too. It discourages weed growth, prevents erosion, and improves your soil all at the same time. Remember the golden rule of gardening: the best results come from feeding the soil and not the plants! Plant roots draw the nutrients they need from healthy soil. A lot of home gardeners forget this and try to force-feed their plants rather than create the conditions in which they can thrive. Plants are a lot like us humans; they need the right environment if they're going to grow and flourish.

Most gardening experts recommend that you spread a layer of mulch between 75 and 100 millimetres or 3 to 4 inches. You don't want a permanent ring of moisture around a plant, so be sure to leave a gap around its stem or trunk. If you don't, fungi may form around the stem and cause rotting. Watering your mulch with seaweed fertiliser once or twice every season will help build the health of the soil even further.

Q&A *'What is the best type of mulch to use in my garden?' asks Damien from Tasmania.*

There are many different kinds of mulch available and it's important to choose the right one for you. Talk to your local nursery about the soil in your area and its general moisture content before making a decision. Drier, lighter mulches work well with wet, heavy clay soils; sandy soils need something a bit heavier. You'll also need to think about what you're intending to plant. There's nothing to stop you using different types of mulches in different parts of the garden to create a diverse environment, therefore a variety of mulches may be needed. There are three basic types: straw mulches, woodchip mulches and homemade mulches.

Straw mulch Pea straw, sugar cane straw and lucerne straw are some of the most common straw-based mulches, and the most desirable, as they break down over time and add rich organic matter to your soil, helping to stimulate the growth of important bacteria. They also help the soil retain a high level of moisture and keep it nice and cool for greater micro-organism activity and protection for the plant roots. On average, the straws will need to be topped up every year.

Woodchip mulch Woodchip mulches are made from raw materials such as tea-tree and tan bark woodchips. Woodchip mulches don't have to be topped up as frequently as straw mulch, but they don't encourage as much good bacteria growth in the soil. On the other hand, they do promote the growth of important fungi. Woodchip mulches need to be topped up every two years or so.

Homemade mulch It's very easy to create your own mulch for the garden with leftover vegetable scraps from the kitchen, and it's cheaper than buying it. Mix the composted scraps with dry material such as leaves or old newspapers – the dry matter will give the compost more bulk and substance. Let the mixture break down, then allow it to settle by spreading it on one area of the garden before applying it throughout the whole area.

It will be ready for use when the consistency becomes slightly airy. It should not be too wet, too heavy, or too dry. The material should feel moist, but crumble nicely through your fingers when you shuffle it through your hand. If you use compost as mulch, your soil will be full of organic material, able to soak up and retain lots of water.

Be sure to do some research before you mulch, and never take the supplier's claims at face value. Some commercially prepared mulches are better than others. There are a few potential problems to look out for. If you're shopping around for a good price, be careful. A lot of councils supply mulch at low cost or for free, but the raw materials may not have been selected carefully. Some of these mulches may contain unwanted seed and you could soon have weeds growing in your lot.

Tanned bark and other treated mulches may contain chemicals that will leach into the soil and have adverse effects – and the last thing you want is to be putting harmful chemicals into the soil. Another problem with woodchips is that they may have come from a company involved in old-growth logging. Logging old-growth forests can result in a significant loss of forest habitats, placing further pressure on threatened species of plants and animals, increasing soil erosion and reducing water flows in streams and rivers.

Choosing plants

There are a million-and-one water-efficient plants available, but there are a few things you should look at before you decide which ones are right for your garden. Choosing plants adapted to your climate and your soil is incredibly important and can save you tens of thousands of litres of water per year.

Australia's climate varies greatly, so different plants will be suitable in different areas. Soil types vary too, of course. Only hardy greenery will flourish in high-drainage soil like sandy loam. If you are set on planting high-maintenance varieties that need lots of water and nutrients in high-drainage areas, it is best to 'work the soil in' first – tilling, composting, mulching and adding in topsoil over several months to create a more water-efficient environment.

If your soil has a heavy clay base, drainage may be poor. Shallow-rooted plants that require little water may become flooded and be easily blown over.

Native plants

Australia has thousands of beautiful, flowering native plants that require very little water to survive. This is no fluke or coincidence. Australian plants have evolved with the land and adapted to our climate. Most Aussies understand this, but we are still obsessed with English gardens. Hanging onto this look after so many years of drought is like using a tanning bed after learning what the rays do to our skin. It's lovely to walk on soft green turf and literally smell the roses, but we must be aware of the cost of all this greenery.

We need to change the way we look at our gardens and what we perceive as attractive. Australian natives and drought-tolerant exotics can flower in the most spectacular way, especially plants like the protea. By choosing native plants, you'll reduce your water needs and attract a beautiful array of native birds, creating a new atmosphere in your garden. This is of course why we have gardens in the first place: to help us relax as only nature can.

Jason Davenport, Horticulture Manager at Royal Botanic Gardens, Cranbourne, in Victoria, says the native garden is slowly starting to become more stylish.

'We need to educate people about plant variety. There's a real diversity and beauty to Australian plants, and they're really underutilised. I think we've still got a hangover from the seventies when the native garden peaked.'

Davenport says people need to spend a bit of time learning about natives before planting them.

'While many will survive on low water, native plants aren't all water savers. Plants that originate from

Australian rainforests are going to need more water than plants from arid parts of the country.'

He suggests people check out bottle trees, dwarf banksia shrubs and firewheel trees. *Scaevola*, otherwise known as fan-flowers, come in beautiful whites, pinks and mauves. Or why not rip out your roses and plant grevilleas? There's a huge range of hybrid grevilleas available – you can get some fantastic colours.

Davenport says the garden of the future is beautiful. But only a small fraction of the plants that are going to be in it are available at nurseries. Ask the manager of your local nursery to bring in some more natives and tell your friends to do the same!

Having said all of this, it is possible to keep an English-style garden in Australia and be good to the earth. It just means you have to recycle more water. Greywater and rainwater keep the majority of pretty little English plants flowering and flourishing – so you'll need to check out our 'Recycling greywater at home' and 'Rainwater harvesting' chapters. But if you don't want to work hard in the garden, plant the largely native greenery that suits Australian soils and climates.

Water-efficient exotics
Certain exotics are able to endure our extreme conditions here in Australia. 'Mediterranean' varieties

are fantastic for our climate. This doesn't mean plants that grow exclusively around the Mediterranean Sea, but any plant that grows in a similar climate in countries around the world – wherever the summers are long and dry and the winters are short, cool and wet.

As a rule, plants with small leaves are more tolerant of drought conditions. They have less leaf surface area and are therefore less affected by evaporation caused by the sun's rays.

We should look on our water crisis as an opportunity to try new plant species rather than a restriction on what we can do in our gardens. I suggest you check out *Gardening Australia*. The television show, the website and the magazine are all great sources of information.

Plant grouping and garden design

When you design a garden, watering needs are just as important as aesthetics. Grouping plants with similar watering needs together is a great way to start a water-efficient garden. For example, you would plant tough natives with other hardy plants, and keep them separate from exotic flowering species that need more water.

A lot of water is wasted when you put a couple of plants that need a lot of H_2O in among others that don't drink much. Imagine a thirsty fruit tree or bed of annuals in a patch of low-maintenance permanents like established camellias or azaleas. The plants that require the least water soak up the excess, even though they don't require it for survival, and you'll end up pouring more and more water on the area to keep up with the requirements of the thirstiest plants.

If plants are grouped thoughtfully, you can reduce the amount of water a new garden will need by some 50 per cent. It's more difficult in an established garden, but if you keep this principle in mind when you're repositioning plants or putting in new ones, you can gradually begin to make a difference.

Water-efficient lawns

The great Aussie lawn is a thing of the past – or at least, it has suffered huge blows. Lawns drink up more water than other areas of your garden, so if you can manage without great expanses of grass, think about replacing them with paving or planting some ground cover.

If you really must plant some kind of lawn for the kids or the dog to enjoy, do so with care. Please do your homework before purchasing turf and avoid disappointment. I've seen far too many bitter homeowners who have been informed it is now illegal to water their luxurious new rolling green.

The major problem with most lawn is that it will keep drinking, regardless of how much you choose to feed it. The best solution is to choose an extremely hardy and water-efficient strain of grass. Your local nursery or turf grower can offer valuable advice on grasses suitable for your area. Some drought resistant varieties include Sir Walter soft leaf buffalo, Palmetto, Windsor green couch, Dawson creeping bluegrass, and Greenlees Park couch.

Of course if you have a big enough rain tank, you may well be able to keep your lawn alive on rainwater. See the 'Rainwater harvesting' chapter for details.

Q&A *'Can I mow the lawn too often?' asks Ingrid from Perth.* Grass roots grow only as long as the blade poking up above the ground, so if you cut the lawn too short or too often, they never get a chance to go looking deeper in the soil for water. They become reliant on the watering they receive from above the surface. So, the taller the grass, the deeper the root system, and the less often it will need to be watered to survive.

Another problem that is rarely considered is that slashed grass emits carbon as it decomposes, adding to the already massive amounts of carbon in the air that are causing climate change. Lawn lovers make matters worse by burning fuel to run mowers.

Lawnmowers/whipper snippers

Keeping a lawn alive is very difficult given current water restrictions, so the Aussie attachment to the lawnmower is slowly loosening, but there are still a lot of people out there who mow their lawn every weekend in summer. It's really not necessary – or appropriate – to do so. Remember that the shorter the lawn the more water it needs from above to survive, so it is best to mow less. Decomposing grass emits greenhouse gas, making it difficult to keep a completely green lawn.

There are several different types of mowers and snippers on the market these days. Here are my thoughts on the different options available.

Petrol mowers/snippers If you are still using a petrol mower, you should try to limit your mowing. Like cars, petrol mowers emit carbon monoxide into the atmosphere.

Some state and local governments are running rebate schemes that allow you to swap your old petrol mower for a battery-charged machine. In Victoria, for example, consumers are offered $200 to swap over from petrol to a battery-powered mower. Check with your local and state government to see what you may be entitled to.

Rechargeable mowers/snippers Mowers powered by rechargeable electric batteries have become quite

Pet peeves: garden products as status symbols

Over the last few years I have noticed a boom in leaf blowers. On any given Sunday, weekend gardeners across the country can be seen wielding one of these beauties. This is the ultimate invention for the non-tradesman who wants to play jostler for the day. Just look at those leaves move. Feel the power of that clean … mmmm …

In one hour, the average leaf blower emits as much carbon as driving a car 500 kilometres.

Do we really need these things? No. How effective are they? Not very!

I have used a leaf blower. The problem with them (besides the emissions) is that they merely blow the leaves on to the road or someone else's property. The next day they are blown back either by your neighbour's slightly larger and more powerful machine or by the cars that race by.

Why do we use them? Because they're cheap and we just can't resist them when we're floating through the aisles at hardware stores buying all sorts of things we don't really need. We buy them for the same reason we buy anything else: because so-and-so next door has one and having a new toy helps take our minds off things that are actually important ... like the water crisis.

I am a man, I know that tools feel good to use, but I can tell you I am still more satisfied when I use the rake to clean up all those leaves, knowing that they will be the perfect ingredient for the most luscious mulch. Those leaves are better on my garden than on the road. And raking them up takes no longer than using the blower.

popular in the last few years and are now sold in most large hardware stores and mower centres. They are a better alternative to the petrol mower and have been proven to produce less carbon emissions. Still, they are run off electricity, and that electricity is likely to be supplied by a coal-powered station. If you have chosen a green energy supplier, though, you are certainly winning on the lawnmowing front.

Environmental mowers cost around $450 at the usual outlets or garden expos. EnvirOmower is the best known brand, and it seems to have a fair hold on the market, but there are several others. Masport is a New Zealand company that makes a number of electric lawnmowers that are available across Australia. Check out www.masport.co.nz for information on where to buy them. The Australian company Victa also makes an electric mower. Find it online at www.victa.com.au.

The problem with electric mowers is that they don't seem as solid as the traditional petrol mower, and if you have a large yard you may need to stop halfway, recharge the machine, and finish the job the next day.

Hand mowers Many of you will be old enough to remember the circular blade mowers that you push along. They are being manufactured again by certain companies like Masport and Rover Mowers (www.rovermowers.com.au). This is of course the most environmentally friendly way to mow your lawn, especially in urban and suburban areas where there is not a lot of lawn to mow.

◆

Be very aware when you buy any garden product that it may be louder than you think. Every decision carries with it a degree of responsibility! Consider your neighbours' ears before you make the purchase. As a rule, the louder products like lawnmowers, whipper snippers or leaf blowers are, the more environmentally unfriendly they will probably be. Battery-powered mowers are much quieter than petrol motors. Even better still is the blade mower or the common rake, which cannot possibly be heard by anyone more than a couple of metres away.

Irrigation

I have seen the devastated look on new homeowners' faces time and time again over the last few years when I've presented the solemn news that their brand new multi-thousand-dollar irrigation system will become – or already has become – unusable. Spray systems and sprinklers are now banned in most states due to tough water restrictions. It makes me wonder how landscapers and irrigation companies can still talk clients into putting pop-up sprinklers in their lawns to be run off mains water.

I have been saying it for some years now: drip irrigation is the only way to go. And that's true for people who are harvesting rainwater or using recycled water on their garden beds as well.

Drip irrigation delivers the water directly into the soil, allowing it to soak in naturally so the plant roots can feed gradually. If you bury the drip irrigation system under mulch, the watering process becomes even more efficient.

Sprinkler systems in general have a high level of run-off or evaporation, and a lot of water ends up on the leaves of the plants, where it does little good.

Micro-sprays or mist sprays may seem to emit a lot less water but again, 70 per cent of the water they pump out never reaches the roots of the plants. You have to leave them running for long periods and on a regular basis in order to achieve effective watering. If you use a spray system, you should seriously consider turning to drip irrigation. It will save you water, and a lot of money, too.

No matter what system you use, remember: you are better off to give your trees a good soaking every two or three weeks than to water lightly every second day. Deep watering breaks through the soil and creates waterways to the plant roots; repeated light watering doesn't.

One of my favourite take-home tips is from Sydney's Royal Botanic Trust: 'train your plants'. It's possible to 'condition' plants to help them cope with drought by slowly decreasing the amount of water you give them. Their leaf tissue becomes a little tougher, making it easier for the plant to get through dry periods.

Q&A *'When is it best to water my garden?' asks Robby, a computer programmer from Melbourne.*

The best time to water your garden is at night or very early in the morning. This is when evaporation is at its lowest. Believe it or not, evaporation still occurs in the evenings, but you lose only a fraction of the amount that evaporates under direct heat from the sun.

If you are using drip irrigation below a good layer of mulch, you can water efficiently throughout the day – more about drip irrigation later.

One last thing: always turn off your watering system when it's raining!

Hand watering

Watering by hand is very efficient if you do it in a responsible manner. It is currently illegal in some states to hand water without using a trigger device that allows you to control and direct the flow. (Old-fashioned garden hoses pump out approximately 1000 litres per hour.) At the time of writing, it is illegal to water with a hose at all in Queensland. Watering is only allowed by the old-fashioned method of watering can, and you can be fined for even leaving a hose connected to a garden tap. Personally I think this is a great move and I'd like to see these laws put in place across the country.

Using a watering can is of course more time consuming, which means that from now on new plants will be chosen with more care, because you'll really need to put in an effort to keep them alive. It is difficult for people with extensive gardens to water by hand, but that is the price we have to pay for having taken our 'blue gold' for granted for so long. There are solutions: get a large rain tank, recycle your greywater or change to natives.

Q&A *'Is drip irrigation better than hand watering?' asks Melanie of St Kilda.*

Hand watering is a good option, Melanie, because you can direct the water straight to the places that need it the most, but I think drip irrigation is the best option of all. The trick is to choose the right type of drip irrigation for your garden.

Recommended irrigation products

I would never endorse sprays or sprinklers. Here are reviews of the three major drip irrigation systems on the market. Please note: no matter which drip irrigation system you choose, it will always be more effective if you bury it below a layer of mulch.

Netafim drip line Netafim is an Israeli company that manufactures several different types of 'in-line' drip irrigation that seem to work well with clean water.

Netafim tubing comes in rolls. It has tiny inbuilt valves spaced around 400 millimetres apart all the way along the hose. The water drips out at a set pace, determined by the size of the hole in the valve. This type of irrigation is good to use where plants can be spaced out evenly, close to the valves in the drip line – for example, next to a row of hedges or a bed of roses. It can be used for other, less formal gardens, too, but some of the water will go to waste where there are no plants under a dripping valve.

Weepy hose Weepy hose is soft tubing with perforations that allow water to seep into the soil. It's a cheap option, and it's been on the market now for a number of years. There are many different brands available. I really like this product because it's made from recycled truck tires. All recycling appeals to me!

Much like the Netafim, weepy hose rolls out. It has

millions of tiny holes all along the line, so you face the problem again of watering where it may not be needed. You cannot control the watering rates, which is essential for some plants. Another problem with weepy hose is that it goes brittle in the sun and tends to block quite easily.

Homemade systems The beauty of the homemade system is that you can place the drippers wherever there are plants that require water. This is the ultimate in efficiency! And it is always more fun to set up your own system.

The system has four basic components:

- the main irrigation line
- feeder tubes
- barbed joiners
- drippers or 'shrubblers'.

Barbed joiners connect the main irrigation line to individual feeder tubes, which direct water to the specific plant or area requiring watering. The dripper or shrubbler is attached to the end of the feeder tube, near the plant's root zone, to feed it directly.

For the main line, all you'll need is some low-pressure 19-millimetre irrigation polly. ('Polly' is tubing made from polypropylene plastic.) You can find it at your local hardware, garden or plumbing supply store.

It will cost around $30 or $40 for a roll of 50 metres. For the average garden, you'll need about 100 metres. You'll also need about 30 to 40 metres of 4-millimetre soft feeder line.

Then you need to buy the actual drippers. I personally recommend 360-degree shrubblers, which have a cap which can be turned to control the flow rate. I'd suggest you choose the 4-millimetre barbed kind. You'll also need 4-millimetre barbed joiners to match. I recommend the barbs and drippers made by a company called Antelco.

Drippers sometimes become blocked after extended use, so it's best to use taps or valves at the end of the line rather than stops or cable ties, so that the system can be opened and flushed out down the track.

You can adjust the flow of water to suit the needs of individual plants by twisting the head. Another benefit of creating your own system is that you can plug the line if you make a mistake and put a hole in the wrong spot, or if you simply wish to change the garden design by adding or subtracting plants. These plugs are aptly named 'goof plugs'!

The tool you use to make holes in the pipe is just called that: a tool. (At least that's what irrigation stores call them.) If you ask for a 4-millimetre irrigation hole-punch any irrigation or major hardware store will know what you are talking about. The solid orange tools are by far the best, because they have a socket with which to punch in the barbs. You can use your fingers instead, but after you've done about thirty or so, expect blisters!

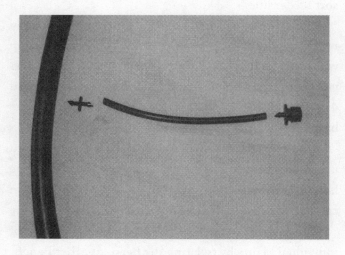

Drip line irrigation: the basic set-up. From left to right: main irrigation line, barbed joiner, feeder tube, 360-degree shrubbler.

If you are having trouble, ask the people who sold you the parts for help. I recommend Reece Plumbing/Irrigation for value and service. Or if you don't have time to take on the project yourself, get a quote from your landscaper, gardener or local handyperson.

In later chapters, we'll look at how you can save even more water in the garden by hooking your irrigation system up to run off recycled greywater or rainwater.

Public green space: more important than ever!

If you walk out to your garden and see crispy brown leaves instead of vibrant, healthy green plants and you're starting to crave the feel of a soft petal, don't despair. Roll up a picnic blanket, pack a book and head off to the nearest botanical gardens. These places are the living breathing souls of our states. They're also offering up lots of inspiration with some innovative waterwise gardening techniques.

Melbourne's Royal Botanic Gardens have cut their water consumption by 60 per cent since the drought took hold and recently opened the Water Saving Garden at their Cranbourne site. The garden is divided into three terraces – the dry terrace, the water-conscious terrace and the waterwise terrace – each maintained under a different watering regime.

Sydney's Royal Botanic Trust is investigating alternative water sources for the Domain and Royal Botanic Gardens including borewater from the Busby's Bore water-recycling project and the installation of rainwater tanks for buildings within the gardens that could provide water to adjacent garden beds.

The Mount Tomah Botanic Garden in the Blue Mountains maintained its collections in 2006 using just over 21 megalitres of collected stormwater run-off – less than half the amount it used in 2005.

At the Mount Annan Botanic Garden in Mt Annan, New South Wales, most of the new gardens have an environmental message. The Bottlebrush Garden is being transformed into a garden of sustainable ideas. It has seats made from recycled milk bottles, garden beds bordered by railway sleepers, car tires used as retaining walls and other resource saving ideas.

Check Australia's botanic gardens out online for more inspiration.

TOP TIPS FOR THE GARDEN

- Use a wetting agent every six months to keep the soil moist and the waterways open.

- Work compost into your soil to keep it healthy and to help it soak up water.

- Spread a layer of mulch in your garden to stop water evaporating from the soil.

- Choose hardy native plants or exotics which can cope with Australia's dry climate.

- Group together plants that have similar watering needs.

- Plant a drought tolerant lawn.

- Cut down on your lawn areas: use pavers or groundcover instead.

- Don't cut the grass too short or too often.

- Fit a trigger nozzle on the end of your hose so that you can control water flow when walking around the garden.

- Water early in the morning or after the sun has gone down to avoid evaporation.

- Water deeply every few weeks – not lightly and often.

- Use a rake, not a garden blower.

- Turn off watering systems when it is raining.

- Get rid of your sprinkler or microspray system and install drip irrigation instead.

- Target root zones when designing an irrigation system or when watering by hand.

- Recycle the greywater from your washing machine or bathroom by throwing it on the garden.

- Use a rainwater tank to collect water for use in the garden.

POOLS, WATER PISTOLS AND RUNNING THROUGH THE SPRINKLER

Every twelfth home in Australia has a pool

When Ballarat went to stage-four water restrictions in 2006, the chairman of the local water authority went on ABC radio to tell people how serious the situation had become. John Barnes, chair of Central Highlands Water, told listeners that at this point, every little bit counts. Kids were told to lay down their water pistols. They are banned under stage four.

Homeowners were urged not to fill their pools, and the water authority is making it hard for people who insist on doing so. Existing residential or commercial pools of 2000 litres or less may be filled using a bucket. The bucket has to be filled at the tap. Larger pools can't be filled at all.

Every twelfth house in this country of paddlers has a pool. Some 500,000 Australian homeowners are now trying to figure out what to do with their swimming pools at the height of the worst drought in a century.

Most of Australia is under some form of water restriction. The rules are different depending on where you live. But the pressure to do the right thing and conserve water during drought is everywhere. Luckily, there are steps you can take to make your pool more waterwise.

First, you can fit your pool with a cover, to prevent evaporation. Be especially careful to cover up your pool on windy days, because evaporation levels increase with wind strength. There's another benefit to using a pool cover: it will keep the water cleaner, and you won't need to add so many chemicals. Putting a deck around your pool rather than grass will also help you limit the amount of water you lose through evaporation.

You can also change the kind of filter you use. Sand filters have to be cleaned with a water-intensive backwashing process that can use more than 10,000 litres a year. If you install a cartridge filter instead, you can clean it with a quick hose down.

If you feel guilty about filling up your pool, there are lots of alternatives. Start by going to the beach and connecting with the natural environment. Or go check out a technically advanced water park.

Whitewater World, a newly opened water-ride park on the Gold Coast, gives the illusion of having water all over its four hectares, but it uses the equivalent of just three Olympic swimming pools to run the ten rides and pools. South-east Queensland's drought forced planners to create a park that could operate under stage-five water restrictions, which could soon be implemented across the region. At full capacity, the Aussie beach-culture-themed Whitewater World can accommodate 4000 people.

On a hot day at home, lots of you will probably have a fond flashback or two of lazy summer days when you were young, running through the sprinkler and screaming with delight. I've spoken to parents who are totally heartbroken by their waterwise kids, who tell them it's a waste of water to do so. Teach them a lesson about fun by putting in a rainwater tank and doing some serious water recycling so you can all play in the sprinkler like you did in the old days.

TOP TIPS FOR YOUR POOL

- Fit your pool with a cover to reduce evaporation.
- Make sure your pool is covered on windy days – evaporation levels increase with wind strength.
- Put a deck around your pool.
- Replace your sand filter with a cartridge filter.
- Go to the beach when you need to cool down.
- Check out a technically advanced water park.
- Install a rainwater tank so that you can run under the sprinkler guilt-free.

GARAGE AND DRIVEWAY

*Do you really need to water
your concrete?*

I see and hear of cases all the time where people still wash their driveways down to get rid of leaves and dirt. I don't get it. I don't know if people can be ignorant of the water crisis anymore. If you have a neighbour or friend who is breaking the restriction laws – watering a driveway down, using a hose to wash the car, or anything of the sort – have a chat with them, and turn the topic to the weather. Ask them if they've heard if rain is forecast, or bring up the latest drought news and just keep talking about it until you think they feel sufficiently guilty.

If they laugh it off, or continue to be irresponsible with water, tell them you have no choice but to let the authorities know. (Maybe word it a little more casually ...) This is not a schoolyard situation. You are not a bad person if you 'dob someone in'. This is our future we are talking about!

If you are driving along and you see someone being a wally with water, don't hesitate to call the authorities. Refer to the back of this book for contact details for the appropriate organisation in your state.

The best thing you can do is to set a good example for others. There are many ways in which we can curb water usage in the driveway or garage, so here are some handy tips.

Washing the car

Cleaning cars, boats, motorbikes and bicycles in driveways consumed a lot of water in the past. It's virtually illegal now, of course. Washing cars with a hose is banned in almost every state and territory, so the question now is whether you keep the laws or not.

You have three main options if you want to wash your car without wasting water: the bucket, the commercial car wash, and the new waterless car-cleaning products.

Bucket

It should really take no more than three or four buckets of water to wash an average-sized car by hand: two for the scrub down and no more than two for the rinse. (Try using the warm-up water you've saved from your shower.) If you use a bucket, you'll use no more than 36 litres for the entire wash – as opposed to the hundreds you'd probably waste with a hose. If you have a water tank, you can use a trigger hose for an efficient rinse when you're done.

Commercial car wash

Look for a car wash in your area that recycles its water. If the nearest one is many miles away, you have to think about the emissions your car will produce if you travel there. They may undo all of your good intentions. Make a sensible decision. If the car wash really is too far away, and you don't have time to wash your car yourself, pay a local child to do it. (I won't lecture on lifestyle ... but many people find it invigorating to clean their own car.)

Waterless car wash products

Last but certainly not least are the cleaning products that require zero water for a beautiful finish. Meguiar's makes a good 'Waterless Car Wash Kit' that is available at Autobarn, Big W and car accessory shops.

> Luke, a plumber from Narre Warren, Victoria, swears by Meguiar's waterless car-wash kit: 'It's just so easy, mate. I've been using it for two years and I have deadset not seen a better finish. It's heaps quicker than a traditional water wash ... and cheap!'

There is a fourth option that rivals them all in terms of time spent and water saved and it is unashamedly endorsed by this author. Simple: *Don't wash your car at all!* May not be for everyone, but it works for me!

The two G's

Let us now discuss those sacred areas, the garage and the garden shed. Traditionally, not a lot of water is wasted in these areas, unless you are cleaning up your workspace – or yourself. It was a ritual for my pop and I that after a morning spent tinkering, welding and hammering (well that was him, I mostly dirtied up the place and caused a nuisance), my nana would yell for 'Schweppes and unch', which in my five-year-old lingo meant morning tea. Schweppes of course meant lemonade and 'unch' was lunch. We would go to scrub up. My grandparents lived on the land and even back then they were aware of the need to use water wisely. So Pop and I would spend some time applying mechanic-grade wash to our hands before scrubbing off the grease under the tap. I guess I'm saying that there's no need to slam on the tap and let the water run galore. Turn on the taps only as needed ... preferably after a good scrub of soap. I know it takes a little water to get a lather up, but you can turn the tap straight off again until you need a rinse. Or even better, just fill the basin once for the whole job.

When it comes to washing the garage or garage floor, take the same approach. If the stains are tough, apply some solvent and let it sit for a while, just as you would when soaking tough stains on your clothes. You won't need so much water to rinse it away when you're done.

Gutter talk

If you have gutters on the garden shed or garage roof, think about installing a small rainwater tank for cleaning these areas and washing your vehicles. You can then have a sustainable garage and driveway area.

There generally isn't much roof space over a shed or carport, so get some advice on what size tank you can put there. (Look at the 'Rainwater harvesting' chapter for the golden rules on working out the ideal tank size for your needs.)

TOP TIPS FOR YOUR GARAGE AND DRIVEWAY

- Sweep your driveway. Don't clean it with a hose.
- Be responsible and talk to your neighbours if you see them wasting water in their yard.
- Call the proper authorities if a neighbour or anyone else persists in wasting water.
- Wash cars and other vehicles with a bucket.
- Use the 'warm-up water' you've saved inside the house to wash the car.
- Use a commercial car wash that uses recycled water and is close by.
- Try a no-water car wash from a car accessory shop.
- Give up washing your car altogether. See how long you can last.
- Don't let the tap run when cleaning your hands.
- Install a small rain tank to suit your shed or garage.
- Let heavy stains such as oil marks soak in solvent before rinsing them away.

RAINWATER HARVESTING

*One millilitre of rain to
one square metre of roof space =
one litre of rainwater*

Imagine turning on your taps to find nothing ... This is a possible reality for country Australians in the near future.

People in rural areas who are not connected to the mains water supply are experts in rainwater harvesting – they have to be, to survive. But as temperatures go up and the drought worsens, many rural people have been forced to buy water to fill their tanks. As the problems increase, prices will only go up. Some are starting to think about upgrading their water tanks. Volume is the key – if you capture as much rain as you can when it does arrive, it will keep you going longer between downpours.

One millilitre of rain to one square metre of roof space = one litre of rainwater. This is the golden rule used to work out how much water you can harvest from your roof. The difficulty of course is predicting the rain!

I run a rainwater tank business, which helps me get an idea of how people in metropolitan areas are reacting to the water crisis – and they are growing more and more concerned. In the summer of 2006/2007, tank orders rose more than 700 per cent, compared to the previous year. As I write, the average wait on a rainwater tank of any kind is approximately fifteen weeks, whereas eighteen months ago, in the summer of 2005, you could have ordered a tank of any size or colour and had it within two weeks.

Up until 1998, the installation of water tanks in metropolitan areas was discouraged by some local councils, even in periods of drought. Some homeowners have told me that after they installed tanks they were actually charged extra rates by water companies – and this happened as recently as 2002.

Now, Australians from coast to coast are being encouraged to install tanks. Government rebates are increasing, giving the public incentive to consider rainwater harvesting as a serious water-saving solution.

People sometimes ask whether installing rainwater tanks is really economical. After all, it costs around $2000 to $3000 to install a tank, and you'll probably only save around $150 on your water bill per year by doing so. That means you won't recoup your investment for around fifteen or twenty years! The good news is that government rebates are available in many parts of Australia to help homeowners shoulder this cost. See the 'Rebate' section at the end of the book for details.

But with or without a rebate, purchasing a tank

is still worthwhile. It's not a question of economics. Installing a rainwater tank to feed your garden or flush the toilet is a moral and ethical choice – and an investment in ours and our children's futures.

Some people ask me whether there's any real point in installing water tanks in inland cities at all. The argument is that while inland cities and towns draw water from rivers, they eventually return the used water back to its source, so collecting water in tanks before discharging it into the river makes little difference, and may even reduce the flow of water downstream.

This argument doesn't take all the facts into account. Because city streets are paved or concreted, and can't absorb rain, heavy downpours can cause a sudden damaging rush of stormwater into local waterways. By catching some of the water, tanks limit the force of that first flush into our creeks and rivers. This is one of the environmental benefits of widespread rainwater harvesting. Tanks trap some of the pollution that would otherwise find its way into our river systems, too.

During hot summers, rainwater tanks also help reduce the demand on the mains water supply. When water restrictions are in place, gardeners with tanks can still water their gardens, but their neighbours using mains water can't. Many people see this independence as a huge plus.

Dismissing the idea of rainwater tanks as uneconomical or impractical is to deny the gravity of the situation we're facing and underestimate the part we all have to play if we are to survive the current water crisis. Even if your tank fills only a few times a year, it's better than nothing at all.

What can I do with my rainwater?

In metropolitan areas facing severe restrictions, most people seem to be opting for rainwater tanks so that they can keep their gardens in good shape or top up their swimming pools, but you can also use rainwater inside the house. Using tank water in the washing machine is increasingly common. You can also use rainwater for bathing and flushing the toilet.

Q&A *'Can I drink the water that I collect in my tanks?' asks Ben from Adelaide.*

Rain water can pick up toxins from the air, pollutants from roof surfaces (which may include heavy metals and chemicals), animal faeces, insects and dirt – none of which you want to drink. Interim results of a major study of rainwater tank quality indicate it is often safe to drink, but far from reliable. The $600,000 study by Brisbane City Council sampled twenty-nine rainwater tanks and found the water met Australian drinking water guidelines in most cases, except where animals and foliage polluted the roof-top water source.

I think it's safest not to drink water from your roof at this point. Stick to the mains supply. There are systems that filter rainwater for city dwellers, but the ones I have seen are designed for use with low-volume tanks and hardly seem worth the cost or effort. Of course it's different for people in the country, where there is less risk of contamination by chemicals and pollutants that run through cities. People in the country have to worry more about their tank water getting tainted with insect larvae, mosquito wrigglers and animal faeces, but all tanks now come with inlet strainers that are designed to address these problems.

Collecting rainwater for drinking and cooking is not recommended in areas affected by airborne pollution from very heavy traffic, industrial activities or agricultural crop dusting or spraying.

Q&A *'Can I shower in rainwater?' asks Hailey from Brisbane.*
It is absolutely fine to connect your rain tank to the shower, but there is one problem: we simply use too much water in the shower, and the average urban tank would empty in a matter of weeks or even days.

As noted above, the quality of water collected in tanks is not as good as mains supplies, so you might not want to wash a baby or someone with a compromised immune system in rainwater. For the rest of us it's not a problem, provided your rainwater tank is well-maintained. The water should be clear and have little taste or smell.

Irrespective of how tank rainwater is used, water quality is dependent on a strict maintenance program. Maintaining tanks isn't particularly onerous, but in practice, most roof catchments and rainwater tanks are poorly kept. That's why the toilet is the perfect place to use the rainwater you collect. Systems connecting rainwater tanks to toilets or washing machines are commonly known as rain-flush systems, and I'll discuss them a little later in this chapter.

Choosing the right tank

It's great to be able to use water from your rain tank in your home and garden, but it's important to remember that you probably won't be able to rely on the tank water alone – it's a supplement to the mains supply. It's difficult to retrofit a tank large enough to supply the needs of an average urban block. In a new home it's different: you can put a huge tank under the ground and store much more water.

As a rule it's best to get the biggest tank you can fit in your yard to make the most of the opportunity. In many cases, a tank with a modular design can be fitted under the eaves or alongside the 'service section' of the house (the side containing air conditioning or heating units), which is a good way to use otherwise dead space. Modular tanks are made up of pieces or modules that can be assembled together to fit a particular space, so they're ideal for pathways or other difficult areas.

It is worth doing a lot of research into the various types of tanks that are available before you buy. Many times have I been to a property to quote or give advice and discovered that the owners have previously installed a tank that is just too small to meet their needs, and, worse still, that they could have easily obtained more volume for approximately the same money.

Have someone like myself to do a free assessment, so that you can get an overview of the many different ways in which you can harvest rainwater. There is no perfect solution when it comes to saving water. Putting in a rain tank always involves compromise: you have to balance cost against available space and aesthetics. The final decision rests with you, but it's good to have the best advice possible before you make a decision.

Most tanks these days are made of four basic materials – polythene, concrete, fibreglass or galvanised steel – and they come in four basic varieties. Let's look at them one by one.

Traditional tanks

Traditional rainwater tanks are the round type we've all seen before. They vary in size and volume and are generally a bit cheaper than other tanks with a similar capacity. Small tanks start at around 300 to 400 litres and the biggest go up to about 50,000 litres. There's no way you'll fit the largest ones in your home, but a 300-litre tank is fairly useless – you'll probably want a medium-sized tank. It's all about balance.

Traditional round tanks are made from all four of the basic materials I listed above, and come in a wide variety of shades. They're available from the big manufacturers such as Nylex, Tankworld and Polymaster, or on order from Reece plumbing and other plumbing suppliers, but I'd recommend going through an installer, because they'll give you a free quote and personalised service.

Waterwall tanks

The Waterwall is a modular tank designed for practicality and effect. They are one of the most popular tanks installed in urban Australia, because they slide nicely under the eaves on the 'service' or unused side of the house. They were invented by a company called Waterwall Solutions based in Melbourne, but they are distributed Australia-wide. Waterwall tanks come in two sizes: 1200 litres and 2400 litres. Several tanks can be linked together to achieve the capacity you require. Since their invention the basic shape has been copied by other companies, but Waterwall remains the best

> Mary, originally from Tasmania and now living in Melbourne, had me install a rainwater tank in her yard. She decided to install a Waterwall tank, because she had limited space down the side of her house.
>
> **Mary:** When I used to live in the country we had tanks, so I'm familiar with them. The advantage of a tank is that you can wash the windscreen of your car or water the vegetable garden, even when there are water restrictions in place. I realise the limitations that city people will soon have to face, so having a tank gives me a sense of satisfaction.
>
> Rainwater is also very nice to drink if you have the appropriate filters. Most people in small country towns in Tasmania are connected to the mains system these days, but the problem is that the mains water comes from creeks – it's untreated, and the water is not nice to drink. So you have a rainwater tank connected to a kitchen sink especially for drinking.

value for money on the market. Check out their website: www.waterwall.com.au.

Bladder tanks

Bladder tanks have only become available over the last few years. For those who've never seen one, they're basically like a big waterbed! The beauty of the bladder is that they can be installed under existing houses that are on stumps or piers. They're perfect for city dwellings on small blocks where there's no space to squeeze a tank in beside the house or in the backyard. Some bladders are designed to sit on the ground (though you may need to level the ground first), while others sit on a support structure just above the ground. The latter style tends to be more expensive and complex to install. The great advantage of the bladder tank is that you may be able to harvest far more rainwater than you would with a traditional tank. The problem is that they're tricky to put in. In other words, hire a plumber who specialises in rainwater harvesting.

Underground tanks

Underground tanks have two major benefits: first, they're hidden away where you can't see them; and second, they can hold huge amounts of water. The downside is that you will have to pay for excavation,

Berryl from Camberwell in Melbourne decided to install two bladder tanks discreetly under her house. She had limited space around the house, but lots of room under it, so bladder tanks were the best option. The tanks were connected to her toilet, shower and washing machine, as well as two garden taps.

Berryl: My reasons for installing the tanks are not all altruistic. The bottom line is that I don't trust the government to supply my water. I just don't want to rely on them for anything. My aim is to become as self-sufficient as possible. The second reason is more ecological. I'm concerned about the future of my children and my grandchildren, because I don't know how bad the water crisis may get; I don't think anyone does. For now, I can use the water on my gardens, but I know we have it there to drink if need be. Combined, my tanks can hold 6500 litres. That may not be enough, but it's a start.

It's definitely worth the money I spent. I think that if you can afford it, you should do something. We may look at putting in more tanks later on. I don't care if they can be seen from the street, because then people will know why I have such a beautiful green garden. The next thing I will be doing is solar panels.

which could prove expensive. In regional areas where space isn't a problem, it's generally easier just to have the tank above ground, but for people in urban areas, installing a large cement tank underground may be the best bet (something in the range of 30,000 litres, perhaps). Cement tanks are often made to measure, so they cost a lot more than a standard tank. You can install certain types of plastic and fibreglass tanks under the ground, but you'll have to do a lot of research if you wish to go down that path. Plastic tanks have been known to collapse under the weight of the earth. Tanks holding more than 15,000 litres may need a pole prop near the centre to stop the roof from collapsing. (Manufacturers should mention this at the point of sale or inquiry.) Plastic technologies are improving all the time, though, so don't let this put you off. Check out the options available before making a decision.

Do I need a pump on my tank?

Many people automatically think they need a pump on their rainwater tank. In some cases you might, but not always. It's a question of how much pressure is required to perform the task at hand. If your block is flat and you're happy hand watering the garden with a bucket, gravity alone may suffice. A small tank stand, about 40 to 60 centimetres high, would give you enough flow to set up a simple drip irrigation system. (The extra height afforded by a stand creates a head of water, allowing you to water your garden without wasting energy running a pump.)

You may need a pump if you want to:

- walk around the garden with a hose, watering by hand
- hook your rain tank up to a long irrigation line or sprinklers
- use the rainwater to fill your toilet's cistern or washing machine.

If you really do need a pump, it's best to do your research before going ahead. Some pumps use a lot of energy, but there are steps you can take to limit the effect it has on the environment.

You could start by contacting your energy provider and switching to green for very little extra cost. If they don't provide green power, ask them why – and then change your provider. See the back of this guide for a list of green energy providers across the country.

Another option is the solar pump. A number of pump manufacturers now make solar-powered pumps of all kinds. Get started by browsing www.solarshop.com.au.

What type of pump do I need?

You will first need to consider how and where you are going to use the water you collect. How far will the water have to run? What type of sprinkler heads do you have?

It is best to set your tank up so that you can hand water with a trigger hose or fit a drip irrigation line. If you really want to run an old irrigation system with pop-up sprinklers and the like, it's important to get a pump powerful enough to do so. (I don't recommend this, as drip irrigation is the most efficient form of watering, and therefore the best way to go, but if you must use a spray system, you might as well use it as efficiently as you can.) Reliable pump manufacturers include Davey, Onga, Lowara, Orange and Leader.

Will a pump make a lot of noise?

Some pumps squeal a little bit under pressure, while others make a low thudding sound. Consider installing a submersible pump that can be lowered into the tank. This will cut the noise and also keep the unit looking very tidy.

The other option is to go for a simple turbine pump with a pressure sensor that cuts the system off automatically when it's not in use. You don't have to turn the switch on and off every time you water; simply turning the tap off or locking the trigger on the hose will cause the pump to shut off. Two popular, cost-effective brands are the Clay-Tech Garden Buddy and Riva-flo by Onga. They're also good for manual toilet-flush systems, which I'll look at shortly.

Rain-flush systems

You can connect your rainwater tank to your toilet or washing machines with what is known as a 'rain-flush' system. These systems can be automatic or manual. Both options have their advantages.

Automatic rain-flush systems

An automatic rain-flush system senses when the rain tank is empty and immediately switches itself back to the mains water system. The major benefit of this system is that you won't need to worry about turning it on and off.

The Davey RainBank system is one of the most popular on the market. A pump and sensor unit retails at around $1000.

A major problem with automatic systems is that the person who installs it will need to isolate the mains connection to the toilet (or the washing machine, or whatever it is that you want to hook up).

This isn't hard if you're building a new home, but if you're retrofitting it might be difficult, as the mains water pipes are inside the walls. If the installer has to start knocking holes in the wall to search for the mains pipe, installing your rain-flush system will be significantly more expensive.

Manual systems

A manual system is just a standard pump, with or without a basic sensor unit, connecting your water tank to the toilet or washing machine. The chief benefit of the manual system is that you can set it up so you can use the tank water in your garden *and* in the laundry *and* the toilet.

Manual systems are easier to retrofit than automatic systems. In the toilet, it's simply a matter of drilling a small hole through the lower toilet wall and adding an extra tap or mini-stop to the cistern. Then you can choose whether to flush the toilet with tank water or mains water. The new tap you install can be left on at all times. (It only runs after you've flushed, when the cistern needs refilling.)

If you want to switch back to mains water, you just turn the tank water tap off and the mains water tap back on. Only one or the other should be left on at one time. It's quite obvious when you need to switch back to mains water: the toilet will fail to flush because the cistern is empty.

I'd recommend that you get some advice from a licensed plumber or tank installer with experience in tank-to-toilet connections.

You can use rain-flush systems in the laundry, too. Just install an additional cold water tap near the washing machine, and hook it up to the rainwater tank via the pump. You'll be able to swap the feed hose to the machine from mains to rain tank whenever you like. To use the rain water, manually unscrew the washing machine hose from the mains water cold tap and connect it to the newly installed rain water tap. If you ever run out of tank water, you can unscrew the hose and connect it to the mains tap again.

Rainwater is generally used for cold water in a retrofit, because there's usually a lot of unnecessary extra plumbing involved in connecting to the hot water unit. You don't really need to wash clothes in hot water anyway – it's a waste of energy. Cold water washes them just as clean.

It's a good idea to have a basic pressure sensor unit on the pump so that you don't have to turn it on and off each time you wish to use the water from the tank. When the newly installed tap is turned off, the sensor will cut the pump off automatically.

Another big advantage of manual systems over automatic systems is the dramatic difference in price. A manual system costs around half the price of an automatic unit. I find that people often prefer the manual system, but if you're building from scratch, an automatic system would make more sense, as they are less work.

First flush diversion

If you are in the market for a rainwater tank, you'll undoubtedly come across the issue of first flush diversion. A first flush diverter is a device installed on a tank to get rid of the first 50 litres or so of water that falls every time it rains.

As discussed earlier in this chapter, rainwater collects pollutants from the roof of your house. If you're drinking the water that falls into your tank, a diverter is essential – even in the country.

If you're just bringing water into the house for use in the shower, toilet or washing machine, or using it on your garden, it's not necessary to install a diverter, but if you have particularly sensitive plants, or use the water on your vege patch, ask an expert whether the type of roof you have or the build-up in your gutters is likely to contain harmful chemicals.

Government rebates

Water tank rebates are generally controlled by state government departments or agencies and passed onto consumers through their water retailer, but in some cases local councils offer a rebate, or top up the state government rebates. I've included information about state government rebates at the back of this book, but I suggest you contact your local council to find out if you can claim any extra money back.

I love to talk about rainwater tanks, so if you have any queries after reading this chapter, please email me at info@thewaterbloke.com.au. And one final note: even though the rainwater in your tank falls free from the sky, it's important to treat it with great care and use it sparingly.

TOP TIPS FOR RAINWATER HARVESTING

- When installing a rainwater tank, have an expert do a free assessment to work out what your needs are.

- Decide what you wish to use the water for and work from there.

- Do plenty of personal research.

- Think about which tank design is right for you and your house.

- Get the biggest tank you can possibly fit in your yard.

- Try putting your tank up on a stand – you might not need a pump if gravity will do the work for you.

- Only use a pump if necessary – they burn energy.

- Try a solar pump, or switch to green energy.

- Connect your tank to your toilet or washing machine with a rain-flush system.

- Use a first flush diverter if you want to drink the water you collect.

- Apply for government rebates if they are available in your state or local council area.
- Use rainwater sparingly to achieve maximum efficiency.
- Visit my website, thewaterbloke.com.au, for further information about rainwater tanks, or email me at info@thewaterbloke.com.au.

If you follow a few commonsense rules, using greywater is perfectly safe

In countries around the world, people quench their thirst with recycled water. Many Australians shudder at the thought, but attitudes are slowly changing as resources become scarce. We may not be quite ready to drink recycled water, but many people are happy to feed it to their gardens and use it to flush their toilets.

What is greywater?

Australia's Environmental Protection Agency defines greywater as wastewater from the washing machine, shower, bath, hand basin and kitchen. Toilet water is called 'black water' or sewage.

The reuse of greywater in the house and garden has started to catch on. As the drought drags on, people are realising that greywater can keep their flowers in bloom. It also has economic benefits. My clients' water bills drop by as much as 50 per cent when their washing machine water is recycled onto the garden. But greywater has had some bad press. Some fear it may pose a health risk. To call it a 'grey area' would be an accurate pun.

I've been educating people about the myths that cloud the greywater issue for years. Until recently, the average person was unaware that greywater could be reused, and others had so many misconceptions about the technology that they'd written the idea off completely.

I've discussed the subject with all sorts of people – from industry experts to budget-conscious homeowners, avid gardeners to politicians. Many seem to have strong opinions, but there's a lot of misinformation out there. I've even met people in the water industry who are in the dark about greywater.

The misunderstanding usually starts with a curious new home builder investigating the various water-recycling systems. They ask their plumber, who says it's just too difficult or expensive a road to go down. It usually doesn't take long before the homeowner shrugs their shoulders and gives up, because it's too hard to find the information about how to do it.

Don't be discouraged if someone tries to tell you it's all too hard – there's a lot more information available today than there was even a few years ago. Companies manufacturing water-recycling systems and giving advice on the issue came on the market far too late and are still catching up to consumers' needs. People like me – the early birds involved with greywater – have been pushing very hard to raise awareness of the issue and improve the technology through experimentation and research. I've learned the hard way, through trial and error, but now I'd like to share what I know with you.

In this chapter, I'll start by discussing how you can use greywater in and around your home and the rules you need to follow to do it safely. Then I'll look at the various ways you can collect greywater for reuse.

Health and safety rules

Only two generations ago, Australians routinely recycled their greywater onto vegetable patches and didn't think twice about it. Recycling and reusing were habits that came naturally then. If you follow a few commonsense rules, using greywater is perfectly safe.

Collecting greywater

You should only recycle water from the bath, shower, hand basins and washing machine. Never use greywater that might be contaminated by faecal matter. For example, don't recycle water that you've used to wash nappies.

Kitchen water should only be used when a full treatment system is in place, because it contains food scraps and insoluble fats that build up in the soil, affecting your plants' ability to receive water. (I'll explain the different types of treatment systems later in this chapter.) Kitchen water is also a potential source of harmful bacteria.

Don't allow children or pets to play with greywater, and always wash your hands after handling it. Wear gloves as an additional precaution.

Finally, never store greywater for more than twenty-four hours without treatment, and stop using it immediately if it smells bad or seems to be harming your plants.

Q&A *'I have heard that there are bugs in greywater that can make you sick. Is that true?' asks Yoshi.*

Greywater contains organic matter. Clothing stained with food, dirt, blood and faecal matter is rinsed in washing machines and the water is then recycled. Fats and food scraps from kitchen waste also make their way into greywater. This organic matter naturally contains bacteria and therefore should not be ingested. If you store the water, the bacteria can build to levels that may be harmful. E. coli and Hepatitis A can grow in greywater. Ingesting E. coli commonly causes gastrointestinal pain, but in extreme cases, it can be fatal. That is why the EPA has put rules in place to make it clear that greywater should not be stored.

If you use greywater properly, you will encounter only trace amounts of bacteria, rarely enough to make you sick. I myself must have ingested litres of greywater over the years in the course of my work. (When servicing systems I've occasionally used my mouth to blow out blockages. Sometimes I've been on the receiving end of a backflow of greywater and unfortunately swallowed it. *Never* do this – greywater shouldn't go anywhere near your mouth!)

Bacteria occur naturally; they are part of any ecosystem. Bacteria cover our skin and infest our inner bodies; they're in the soil in our gardens and in our plants' root systems. All of these bacteria actually have a function, and work together or against each other to create balanced and diverse living environments.

Using greywater on your garden

Most people who choose to install greywater systems use the water they collect in the garden. It's a great way to beat water restrictions and natural drought. There's a bonus: the extra nutrients in greywater give your garden a real boost. Azaleas, camellias, roses, fruit trees and tomatoes are just some of the plants that thrive on recycled water. When you make the switch, you'll often note a higher yield or number of blooms. Once again, though, there are some basic rules you must follow.

When using greywater in your garden, you should run it through a drip irrigation system, making sure the main line is buried beneath a good layer of mulch. In some states, environmental authorities require untreated greywater to be delivered 75–100 millimetres below the surface of the garden bed, to eliminate any health risk. This is easy to do: just build a thick layer of mulch over the irrigation. Mulching will help the water soak into the soil and also prevent evaporation.

Try not to water the same area every time. Spread the water evenly, or rotate between several different areas. And don't water your vegies if you're planning to eat them raw.

You don't want greywater flowing off your property and into the stormwater system or onto your neighbours' property (it's against EPA regulations), so don't use too much. Give the soil and plants a chance to absorb it, and don't use it on the garden during wet periods.

Finally, stop using greywater straight away if it seems to be harming your plants.

Q&A *'Can the chemicals in my soap or laundry detergent harm my plants?' asks Dorothy, a keen gardener living in Dulwich Hill, NSW.*

This is the most common question I've been asked by people over the years, regardless of demographic, age or background. Before you funnel your used bathwater onto the garden, you need to look at the soaps and detergents you're using, particularly if you're pumping the water straight onto the garden without treating it. If you have a full treatment system, there's not as much need for concern, but I always suggest you choose products that are low in chemicals anyway – the more chemicals you use, the harder the system has to work to clean them out.

The soaps we use every day in the shower contain fats that can build up in the soil and make it hard for plants to drink up moisture. Using an organic wetting agent every six months or so will help. (Well-known brands include SaturAid, MoisturAid and Hydraflo2. Please see the 'Garden' chapter for more information about wetting agents.)

Be careful when you shop for laundry detergent, too. Too much sodium, chlorine or phosphorous will have a detrimental effect on your plants and your soil, throwing the pH level out of whack. If you continue to use the wrong detergents, plants will start to look unhealthy and in some cases you may kill them.

Salts in detergents are just fillers – bulking agents – and do absolutely nothing to clean the lining of your pants. They're unnecessary, but many companies continue to use them because they're cheap. Luckily

Soil pH levels

The PH level of your soil tells you if it's too acidic or alkaline.

The pH scale goes from 1 to 14. A reading of 1–6 is acidic and a reading of 8–14 is alkaline or 'basic'. Seven is a neutral reading. Your soil reading should ideally be as close to neutral as possible, so that the soil will be able to support a diverse array of good bacteria and plant life.

Greywater that contains high amounts of phosphorous and sodium can raise the pH level of your soil and lead to an iron deficiency. If you notice young leaves on your plants turning yellow or their veins turning green, you should test the soil's pH. Simple, cheap testing kits are available at nurseries, pool stores and in the garden section at major supermarkets. If you have a problem, a treatment such as sulphate of iron or chelated iron will help. These treatments usually come in powder form and need to be

> dissolved in water. Follow the instructions on the product packaging, as they vary. If the leaves are yellowing and the plant is young, it is usually best to spray the treatment over the foliage, but if the plant is mature, feeding the roots directly is the best thing to do.
>
> Prevention is better than cure, though, and if you use garden-safe detergents in the first place, you shouldn't run into these problems.

there are new brands on the market now such as Green Care and Earth Choice that make products suitable for your garden. Even more surprising ... they are cheap! Check our list of favourite products at the back of the book for further details.

Detergents with the 'NP' (No Phosphates) label are phosphate-free. Detergents labelled with the symbol 'P' (Low Phosphates) comply with the Australian standards for garden-safe products. I don't think the current standards are high enough, so these labels aren't much use. Even if a product is low in phosphates, or has no phosphates at all, it may be loaded with sodium or chloride.

As a general rule, concentrates are better than non-concentrate powders, and liquid detergent concentrates are better still. For more information on sodium and phosphate levels in particular detergent brands, log onto www.lanfaxlabs.com.au.

Q&A *'Can I still use whiteners and softeners?' asks Rusty from Charleville in Queensland.*

Bleach products and disinfectants containing things like eucalyptus or tea-tree oils will kill the organisms in your soil, ruining its health. Avoid them!

If your greywater system runs directly from your washing machine to the garden, but you really need to bleach your whites, here's how you can get away with it. If you have a simple set-up with a basic rubber nozzle coming out of a pipe outside the house, just remove the nozzle so the used water goes down the drain into the sewage system, and not on your garden.

If you have a slightly more sophisticated system with a diversion switch, all you need to do is turn the diverter switch back to the drain.

> Robin is an extremely passionate gardener from Blackburn in Melbourne's south-east. In January 2007 she had me install a greywater system in her home that will help her recycle more than 50,000 litres of water per year, leaving her with little (if any) need to use mains water on her garden.
>
> **Robin:** I'd been thinking about it for ages. This garden has become a kind of obsession for me. I knew that stage-four restrictions were coming and I didn't want to lose it. I simply wasn't prepared to let it go without a battle.
>
> We had a couple of 40-degree days in a row over summer and I decided that I had to do something. I guess I felt in a way that the greywater system would pay for itself. It all has to help. As long as I know the water is going onto the garden, I'm happy.

Q&A *'I want to use bleach to clean my bathroom. What do I do, Craig?'*

This is a question Amy asked me. She has a greywater system hooked up to her shower and washing machine.

Jillian lives in Kew, Melbourne, with her husband. Like Robin, she has just had a greywater system installed, but hers is slightly different: we focused on the shower water that came from the house, feeding it into an irrigation system for the garden.

Jillian: First we looked at rainwater tanks. Then we realised that the amount of greywater we produced was significantly more than we could catch in a small rain tank. I'm away working a lot, but my husband is usually here and showers every day, which means that the garden gets water each day automatically.

I'm glad we've done it. You feel like you are doing your bit. We should really have done something six months ago. It was actually difficult to get much information on recycling greywater. The government should make it compulsory to install underground rainwater tanks and greywater systems in any new house or renovation.

The greywater flows into a tank and is pumped out onto her garden. She wanted to clean the shower with a bleach-based product. (I don't use bleach at all, but I can't stop other people from using it – even Amy. But if you pump bleach out on your garden, you'll be in for a rude shock when the garden dies!)

All Amy has to do is turn off the pump before scrubbing out the shower. When the pump isn't connected to a power source, the water goes into the pump box and overflows into the drain. The chemicals go down the drain too – not onto the garden.

All greywater systems must, by law, have a fail-safe mechanism of this kind. If something goes wrong, whether it be blocked pipes, natural flooding or an electrical malfunction, you need to have a fail-safe mechanism in place to divert the greywater immediately back into the sewage system so that it won't overflow or pool on your property. The most common fail-safe device is a simple overflow or diversion switch. This should be automatic, in case you're not at home when the problem occurs, but you'll need a manual override option, too, if you want to use bleach, hair dye or other chemicals in the bathroom.

Using greywater in your house

Treated greywater can be used inside for just about any purpose ... except drinking. You can also use untreated greywater in the home, but it's a bit impractical, as EPA guidelines specify that you can't store the untreated water for more than twenty-four hours, so it's best just to pump it out onto your garden. If you want to use untreated water in the house, toilet flushing would be the only safe application.

Q&A *'I want to use greywater to flush back into my toilets. How do I do it and how much does it cost?' asks Wei Ching from Marrickville, NSW.*

You need to enlist the services of a licensed plumber. The plumber will check whether there is access to the outlet pipes from your shower or washing machine, and then devise a way to divert water from them into a sump pump connected to a storage tank. You could then feed the greywater into your toilet using either gravity or a pump, whichever was most suitable.

This is quite a technical thing to do and should really only be attempted by a licensed plumber, but if you have your own property and are handy, you could give it a go. The cost will depend on how much of the work you do yourself and whether your home is new or not.

Untreated greywater can only be stored for twenty-four hours, so you would have to empty the storage tank every day to comply with the laws set by the EPA – or treat the water so that it can be stored safely and used when needed. Systems such as the Greymate,

EcoCare, and H2grO units have built-in timers that automatically pump out any greywater left in the sump box after twenty-four hours.

Q&A *'Will using greywater in the house dirty the toilet bowl?' asks Eric, a graphic designer who likes to keep his house looking schmick.*

If you treat your greywater, it should be just as clean as mains water, and you won't have any problems, but if you use untreated greywater to flush your toilet, it will make the cistern and the bowl dirtier than usual. Greywater contains small amounts of lint, detergents and soap that build up quite quickly. This just means you'll have to clean the toilet slightly more frequently.

Buckets

There are a range of quite sophisticated ways to collect greywater for reuse, but before I talk about the various systems available, I'd like to have a quick look at the cheapest, simplest way to recycle the greywater from your shower or washing machine. That's right – buckets.

When showering, use rectangular buckets and actually stand in them. Most of the water will end up in the buckets, and when they're full, it's probably time to stop showering. Two 15-litre buckets is the way to go.

Recycle the water in a practical way. If you live in the city and have no garden, use the shower water to fill up the washing machine to wash your clothes, then use that water to flush the toilet ... Or dump the shower and washing machine water onto your pot plants. All with buckets!

> **Craig's favourite water-saving tip:** *Remove the lid of your cistern, turn the mains connection to your toilet off and fill the tank with the buckets of water you've saved from your shower or washing machine. The only thing you need is the ability not to worry about the look of a cistern with its lid off and a few buckets in the shower. Simple!*
>
> *I live in an inner city apartment and have no garden, so this is my favourite water-saving trick. It saves me over 60 litres of water per day. I haven't used mains water to flush my toilet in over a year.*

You can buy plastic buckets for a couple of dollars, so even if you're on a limited budget, there's no excuse – you can start saving water straight away. If you have a little more money to spend, though, you should consider installing a greywater system.

Greywater systems

There are two major types of greywater recycling systems. *Diversion* systems disperse the water automatically on your garden from either the bathroom or laundry. They do not store the water. *Treatment* systems store water and treat it before it is reused.

It is cheap and easy for homeowners (especially those in urban areas) to install or create a simple diversion system to reuse greywater in their gardens. I usually suggest that people use rainwater inside the house and greywater outside of the house, but this is certainly not a hard and fast rule. If you install a full treatment system you can safely use greywater in your house. The only catch is that treatment systems tend to be more complex and more expensive than diversion systems, so they won't suit everyone.

If you are building a home from scratch, it's fairly easy these days to incorporate water-recycling systems into the building and planning process, but if you're retrofitting an existing house, you'll have to work with what you have. Both diversion and treatment systems range in price and quality. Systems may not be as expensive as you think to install. Call an expert to look at your needs and set out all the possibilities available to you. Contrary to popular belief, there is always a solution of some sort. If you are not happy with one assessment, get another one. Make sure you consider all the options.

Diversion systems

The way the plumbing in your house is set up will dictate whether you should divert water from your washing machine or your shower and bath. The washing machine is generally the easiest.

There are two basic types of diversion systems: gravity systems and submersible pump systems.

Gravity systems If you go for a gravity system, you'll use the pump on your washing machine to run water into a small 'header' tank that sits either on a stand or on the wall and feeds the garden via a drip irrigation system.

You can also hook a second-storey shower up to this kind of system. In this case there's no need for a pump; gravity will carry the shower water into the header tank.

The other option is to plug a rubber bit into the pipe which usually carries greywater from your laundry or bathroom out of the house and into the sewerage system. You can then divert the water onto your garden or lawn via a hose. These rubber fittings are available from all major hardware and plumbing stores for around $15. This set-up doesn't provide much head pressure, and the hose may only work on a downslope.

Submersible pump systems A submersible pump system allows you to divert the greywater from your shower or your washing machine or both, as long as you can access the pipes. The pump sits in a box pumping out the water as it comes in. Some systems allow you to pump the water straight into an irrigation system or into a small tank. You can create a gravity feed from the tank onto the garden, again via drip irrigation.

These systems are perfect for a house on stumps. The plumber has lots of room to get in under the house and work on the drain pipes, and it's easy to set up the graded fall needed to fill the sump pump box. They're also suitable wherever the outlet pipes from the shower and the washing machine can be identified outside the house and directed to the sump pump.

Treatment systems

Treatment systems are generally used only in rural areas or when the house is being built from scratch, because they tend to require a lot of digging and specific plumbing infrastructure.

There are a few different types of treatment system on the market, and they vary in price. Most of us will be familiar with septic tanks; until very recently, they were common in the suburbs, and they are still used widely in the country. Aerated treatment systems are another alternative; they destroy bacteria by forcing oxygen into wastewater. Other treatment systems use toxic chemicals such as chlorine and bleach, but I suggest that you avoid these. Let's look at the different types of treatment in more detail:

Septic systems Septic tanks are commonly used in places where there is no sewage infrastructure. They're made up of two parts: an underground double tank system built of concrete, fibreglass or polyethylene and a series of 'drainfields'.

Solids and liquids from the household wastewater separate in the tank, and the heavy solids settle at the bottom. Bacterial action produces gases, causing lighter solids, fats and greases to rise to the top of the fluid, forming a layer of scum. The solids are pumped out every couple of years or so and the fluids are drained off into the drainfields, where natural bacteria take care of the final stage of the treatment.

Drainfields are drains dug out from the septic system into paddocks, like small canal systems to disperse the water. Shallow-rooted plants or lawns are the final filter. You can always tell where drainfields are if you're out in the country, because the grass grows lush in these areas and farm animals gather around to feed on it.

Septic tanks may need to be cleaned out every three to five years to prevent the drainfields from clogging up.

If you don't do this, you'll eventually notice drainage slowing down and the sewage could even back up into the house. Some systems contain a pump to expel the fluid.

Aerated systems Aerated wastewater treatment systems can be used to treat sewage and greywater, or just greywater alone. Wastewater is filtered through layers of sand and gravel, then exposed to micro-organisms, which help to break down harmful bacteria. The water is then aerated and finally disinfected using chlorine or ultraviolet light before it is pumped out onto the garden.

The two major brands are Envirocycle and BioCycle. The systems cost around $10,000 and you can expect to pay another $500 a year in power and maintenance costs. These systems are best suited to houses in regional areas, where the blocks are larger, but they might be a good alternative if you are building a new house in an urban or suburban area.

Reed bed/filtration systems The final option is a reedbed or filtration system. Greywater is expelled from the house and into a sock or bag filter that sits abreast a drum or small tank. This first filter catches larger solids such as hair and lint. The water then filters into the tank, where it passes through a layer of fine sand, then another of fine gravel, and finally a layer of coarse gravel. The filtered water runs out of the bottom of the tank into a reed bed in a sealed planter box of soil and gravel. Once the water passes through the reed bed it can be collected for safe reuse via a tap placed at the bottom of the planter box.

Commercial versus DIY

These days there are more and more systems flooding the market and they vary greatly in cost. A high-end, full treatment system is always a great way to go, though, as I noted above, they can be hard to retrofit and the price will put some people off. The lower-end systems are really only intended for the garden.

The benefit of rigging up a little home system for yourself means that you'll be able to achieve results extremely cheaply, though you'll have to shop around for the best prices on parts. Now that you understand the basics, with a bit of research, you could easily put one together. You should note, though, you're legally required to have a licensed plumber hook up the actual overflow to your sewage. He or she will be able to supply you with a plumber's certificate, and in some states, this will make you eligible for a government rebate.

Rules and regulations

Environmental, health and/or water authorities in each state regulate the use of greywater and determine which recycling systems are approved. Local councils may also have special requirements for the installation of greywater systems. Check with the authorities in your area before you get into greywater recycling.

Rebates for greywater systems

Queenslanders, Victorians and Western Australians will be pleased to know that if they install a greywater system and have documentation to prove it, they will receive a rebate of up to $500 from their water company in the form of a credit on their water bill. All they need as proof is the plumber's certificate and the supplier's invoice or receipt.

At the time of publication, the other states and territories don't offer rebates for the installation of greywater systems, but this may change, so check with your water authority to find out.

TOP TIPS FOR RECYCLING GREYWATER AT HOME

- Recycle greywater only from the bath, shower, hand basins and washing machine, unless you have a treatment system.

- Don't use greywater that has been contaminated by faecal matter. For example, don't recycle water that's left over from washing nappies.

- Don't store greywater for more than twenty-four hours.

- Wash your hands after you handle greywater.

- Don't let children or pets play with greywater.

- When you water the garden with greywater, rotate the areas you use it on.

- Don't use greywater on your vegetables if you plan to eat them raw.

- Use only as much greywater as the soil and plants can absorb.

- Don't allow greywater to flow off the property or into the stormwater system.

- Don't use greywater on the garden during wet periods.

- Stop using greywater if it smells bad or seems to be harming the plants.

- Use buckets to collect shower water and use it to fill the washing machine or flush the toilet.

- Consider installing a greywater system. Investigate all the options available to you, and seek expert advice.

- If you're unhappy with the first assessment you're given, get another one.

- If you're eligible, claim a rebate for installing a greywater system.

PART 3
THE BIG PICTURE

WHERE DOES OUR WATER COME FROM?

Water in Australia comes from three sources: groundwater, dam water and recycled water

Country people tend to be aware of where their water comes from, because shortages affect them directly. It's different in the city. Some town folk know where their water comes from, and even stay in touch with the levels of their dams, but I would say that most couldn't even name the dam that makes it possible for them to shower!

It's easier to remember to save water if you know where it comes from and understand that it's a limited resource, so in this chapter I'll discuss Australia's water sources, and talk about the different types of sources and how they are managed.

What types of water do we use?

In Australia, our water comes from four major sources: groundwater, surface water, rainwater and recycled water. Three of them come initially from the sky.

Rainwater is collected on our roofs and in tanks, and recycled water is the water we collect and reuse. For people in the country, rainwater tanks are a major source of water for the house, whereas recycled water only counts for a small amount of household water use. See the 'Recycling greywater at home' and 'Rainwater harvesting' chapters for more info on these key sources. In this chapter, we're going to talk about groundwater and surface water – the two major sources of water for the average urban home.

Surface water

Most Australian cities rely on surface water for survival. All water starts out as surface water. It's the stuff that falls from the sky and fills our reservoirs. In periods of long drought, the surface water that feeds into our dams is naturally scarce, but climate change has made this problem worse. Rain now falls in different areas; and the old catchment areas don't receive as much as they used to. The same amount of water is still falling; it's just that it's hard to predict where it will be falling in ten or even five years, so choosing locations for new dams is becoming increasingly difficult. Suitable sites are rare, particularly as dams have a huge enviromental impact.

I believe the money we might potentially invest in new dams over the next few years would be better spent on installing a rainwater tank in every home in every

Australian city. (And not just because I run a rainwater tank installation business.) We'd no longer have to rely on surface water from one catchment area. In effect, we'd all be collecting surface water before it hit the surface, supplementing our existing catchments.

Groundwater

Groundwater is tapped from pockets beneath the earth's surface. It's surface water that's seeped through the surface to form underground reservoirs called aquifers. Aquifers are mined both by private individuals and in large-scale public or commercial projects.

There are two types of aquifers. The first is the shallow or 'unconfined' aquifer. The upper level of these aquifers is known as 'the water table'. The second type of aquifer occurs much deeper below the surface and is known as 'artesian' or 'confined', meaning that the water is under pressure between layers of rock.

Residents of Perth rely on groundwater for almost 60 per cent of their domestic supply. The main confined aquifer mined for domestic use is around 1000 metres thick, while the shallower, unconfined aquifer is about 50 metres thick.

Adelaide and Canberra also use groundwater, but not as extensively as Perth. Our other cities rely on dam water, because they don't have aquifers large enough or accessible enough to tap.

However, mining groundwater has some serious environmental consequences. Underground reservoirs have been created over hundreds and thousands of years, and help to cool the earth beneath the surface. If we destroy the water table through mining aquifers, we risk accelerating global warming.

There's also the risk of subsidence. Mexico City was built above an aquifer. The city's inhabitants have been mining their underground water for over a century now. Environmental scientists estimate that the city has sunk more than nine metres since they started – that's the height of a three-storey building! A scary thought, given Perth's heavy reliance on groundwater; however, Mexico City is built directly on top of its aquifers. Perth's aquifers aren't right under the city, luckily. The surrounding environment is at a bigger risk than the city itself.

Some researchers are looking at the artificial replenishment of aquifers as a solution to this problem. This process is known as 'managed aquifer recharge'. Reservoirs are refilled under controlled conditions using water from various sources, including recycled water. There are major recharge projects happening in Canberra and Adelaide, but the jury is still out on the results. The effects on water quality will become clear over the long term.

The obvious problem is that no matter how you treat the recycled water or purify it, it is never going to be of the same quality as the water that has been removed in the first place. We cannot yet be sure how this will affect the environment in the long run.

Pollution is yet another concern. There are more than 80,000 toxic dumping sites under the surface of Australian soil. These sites are only growing in size and number. Pollutants have been seeping from them into our valuable aquifers. We should be spending a lot more money on researching ways to ensure groundwater is kept at 'A Class' or drinking standard. We should also be trying to cut down on the amount of toxic substances we generate by recycling more, using less packaging and coming up with more intelligent modes of disposal.

Where does your water come from?

If you've never seen the place where your water comes from, why don't you go and check it out? It will really change the way you understand water, and make you stop and think before you turn on the tap.

There isn't enough room in a book like this to give a state-by-state summary of the major water sources around the country – there are just too many of them

– so if you'd like to find out exactly where your water comes from, I suggest you contact your local water authority. You'll find the contact details you need at the back of the book.

The Big One

The biggest groundwater supply in the country and one of the biggest in the world is known as the Great Artesian Basin. The basin was formed between 100 and 250 million years ago, and underlies approximately one-fifth of the continent. It works its way from arid and semi-arid areas of Queensland through New South Wales and South Australia to the Northern Territory. It stretches all the way from the Great Dividing Range to Lake Eyre, covering an area of 1,711,000 square kilometres.

The basin is said to contain 64,900 million megalitres – that's 64,900,000,000,000,000 litres!

There are over 1500 bores throughout the basin, releasing a total of 1500 megalitres of water per day.

Source: www.nrw.qld.gov.au/water/gab

The Great Artesian Basin

THE MURRAY–DARLING BASIN

The famous Murray River Cod and Murray River Red Gums are dying at an alarming rate

The Murray–Darling Basin is known as Australia's 'Food Bowl', supplying up to 50 per cent of the nation's food. The water from the basin travels through four different states. It provides the majority of South Australia's water and is used to irrigate huge areas in Victoria, New South Wales and western parts of Queensland.

The health of the Murray has been suffering for years. The situation was serious enough to prompt a special summit on the river's health in November 2006. At the summit, senior officials presented a report stating that inflows into the Murray were 60 per cent less than the previous minimum.

We can't just blame it on the drought, though. Three-quarters of the trickle that remains is siphoned off for irrigation of our crops. The states are responsible for allocating water to agriculture and industry, and the downstream states inevitably suffer – along with the river itself – when the upstream states are too generous in their allocations. This can't go on.

Several species of fauna and flora are suffering

massively from the result of low water flow through the river itself. The river is home to over eighteen species of native fish that are particular to the area, along with many species of native trees. The famous Murray River Cod and Murray River Red Gums are dying at an alarming rate. The loss of these iconic species would not just be a tragedy in its own right, it would also have a terrible impact on tourism in the area, an important source of income for local communities.

Low flows mean that the river does not get the flushing it needs from time to time to stay healthy, and stagnant, shallow waters lead to pollution problems. Effluent from industry surrounding the Murray compounds the problem, making it even more difficult for native fish species such as the cod to survive, though introduced fish such as carp seem to thrive in these polluted conditions.

Any local will tell you about the carp problem in the river. Carp were introduced into Australia from Germany in the 1850s by homesick settlers. A number of fish entered the river due to major flooding in the 1960s and, like the cane toad in northern regions, they are now a serious threat to our prized local species. Fisherman have a policy of killing them before throwing them back in to the river. (For more information visit: www.savethemurray.com.au.)

Probably the biggest issue the river faces are the open irrigation channels that feed the basin's farming regions. For decades governments have been talking about converting these channels to pipelines, but it's only recently that we've started to see any action. Enclosing these irrigation channels will save millions of litres of water per year in evaporation.

In late January 2007, the Howard Government announced a $10 billion funding package for the Murray–Darling Basin. The enormous sum pledged will be used to upgrade and replace irrigation systems, create new infrastructure, carry out research into water-saving agricultural techniques, and implement new practices. However, the prime minister has stipulated that the $10 billion plan is conditional upon the states handing the federal government control of the Murray.

The federal government would then take over the management of the area's water resources, relieving the state governments of the responsibility. The federal government has repeatedly scolded the states for doing a bad job when it comes to water conservation. Prime Minister John Howard has said the states are to blame for the current crisis, with their over-allocations to irrigators being the main problem.

The creation of a single body to manage Australia's most important watercourse, though essential, actually requires a change in our constitution. Under the current law, the Commonwealth does not have authority over the states' use of water for crop irrigation. This piece of red tape could be tough to cut.

The states are worried their farmers will be forced to sell their water rights back to the government and that cotton and rice growers will be unfairly targeted. The government has said it's 'not inconceivable' that there might be forced buyouts. In fact, $3 billion has been set aside for the buyback of water rights.

Despite these worries, Queensland, New South Wales and South Australia have all agreed to the plan. As this book goes to print, only the Bracks Government in Victoria is holding out, describing the proposed scheme as 'half-baked'.

I hope the states unite on this. It would be the biggest move ever in the history of water management and, I might add, just in the nick of time.

Sceptics have suggested that the pledge is only a campaign ploy in the lead-up to the 2007 federal election, which may be true in part, but it does speak to the gravity of the problem we face. The $10 billion worth of taxpayers' money the government proposes to spend should restore some hope and teach us to conserve the greatest resource that we have: our land and environment.

DAMMED IF THEY DO

Even in good times, Australian dams must capture roughly six times more water than dams in Europe to provide the same yield

Many people I've spoken to while doing research for this book say, 'We need to hurry up and build some more dams.' As though if we build the dams, the rain will come. Well, it doesn't quite work that way!

Dam building has long been the way we've gone about harvesting more water in Australia, so it's not surprising that many Aussies are looking to dams as a solution now. A *Herald Sun*/Galaxy opinion poll taken in October showed that 74 per cent of Victorians supported the construction of a new dam to tackle the state's water crisis. Unfortunately, the solution isn't that simple.

In the last twenty years, urban Australia has constructed many dams. Perth built eight and none of them filled up with water. (Perth's rainfall has declined by around 20 per cent over the last decade, and the flow of water into its dams has dropped by around two thirds.)

Regional areas are in trouble, too. The three dams that underpin the Murray–Darling Basin's irrigated agricultural industry – accounting for about half of Australia's total farm revenue – are nearly empty.

Even in good times, Australian dams must capture roughly six times more water than dams in Europe to achieve the same yield, because of erratic rainfall and high evaporation. These days the rain's hardly falling and virtually every urban centre has been experiencing a record water shortage.

Sydney, Melbourne, Brisbane and Perth are already using more water than is going into their dams, and supplies are expected to drop by 25 per cent within twenty-five years as a result of climate change.

The reality is that we aren't getting enough rain to fill the dams we already have. So should we be building more? It's a difficult question – especially when you consider that dams are hugely expensive to build and can be hard on the environment. They have been linked to soil erosion and species extinction.

What do the experts say?

Scientists and water experts are divided on the question of building more dams. Amy and I spoke with Ross Young, the executive director of the Water Services Association of Australia, a group of companies that supplies 75 per cent of Australia's water. Here's what he had to say:

'Dams are becoming an increasingly problematic option for Australia. Dams only fill when you've got run-off, and run-off is becoming an exceedingly scarce commodity as our rainfall patterns change.

'Calls for increased dam building lead us to the billion-dollar question: Is what we're experiencing now a taste of things to come, or is it an aberration? We don't know what we will have to live with if we build more dams now. It could be an extraordinary waste of the community's money. That's why diversifying is important. Dams are completely dependent on rain. We need to develop sources of water that are not dependent on rain.'

It's not that Young is anti-dam. He says we need to look at all of the solutions available. Dams could be the answer on large rivers where rainfall is good, such as those up in the tropics: 'In the right climate, in the right spot, a dam might be the way to go.'

But he says most of those sweet spots have already been developed.

'In the context of unreliable rainfall and climate change, dams increasingly represent a high risk source of water.'

Proposals for new dams

Queensland has set aside hundreds of millions of dollars to purchase land from residents to make way for two major reservoirs. Proposals for new dams in regional Victoria have also received some recent press.

Controversy in Queensland

Brisbane is the only major city planning to build new dams at present. As Malcolm Turnbull, the Howard Government's parliamentary secretary for water, has pointed out, Brisbane already has the largest storage capacity relative to demand of any Australian city, yet its water shortage is the most severe.

The state government has warned of massive economic damage to south-east Queensland if new water sources are not found. The lack of new water sources could end up costing south-east Queensland at least $55 billion and perhaps as much as $110 billion by 2020, according to the consultants ACIL Tasman.

The Beattie Government is planning to build a $1.7 billion dam on the Mary River at Traveston Crossing near Gympie. According to Queensland Water Infrastructure (QWI), the proposed Traveston Crossing Dam could provide up to 45 per cent of the additional water required by south-east Queensland by 2050, with anticipated yields of around 70,000 megalitres when stage one of the project is complete, and more than twice that when stage two is finished. Traveston Crossing is the last remaining large dam site in the region. An environmental impact statement is currently being prepared, and will require both state and federal government approval.

The proposal has enraged local landowners, prompting vows from several federal Coalition MPs to block the development. Some residents living around the proposed dam site say the development is their worst nightmare. QWI has reached agreements with the owners of around 50 per cent of the required properties, but many others could face compulsory acquisition of their land if they have not voluntarily sold it before environmental checks are completed.

In October 2006, Deputy Premier Anna Bligh announced less land was required for the dam, and 597 properties would be affected. The figure previously announced had been closer to 1000. Eighteen properties not previously affected were suddenly on the chopping block and sixteen already purchased were to be offered back to their owners. Residents have been speaking out in their local media saying they are sick of being jerked around and living with so much uncertainty.

Nationals senator Barnaby Joyce says he wants Prime Minister John Howard to make a clear statement opposing the Traveston Crossing dam. He has told the media that it will turn out to be nothing but an 'evaporation pond in a bog', providing little water and destroying people's lives. He said he was particularly concerned at the plight of small settlements such as Kandanga, which would be flooded.

The Queensland government has also committed $400 to $500 million towards building another major dam: the Wyaralong, south of Brisbane, which has sparked further controversy.

The proposed dam will be located on Teviot Brook, 14 kilometres north-east of Beaudesert in the Logan River catchment, and should supply up to 21,000 megalitres a year when completed in 2011.

Victoria

The CSIRO forecast in 2005 that in a worst-case scenario Melbourne would lose 35 per cent of water flowing to its dams by 2050. Nevertheless, the National Party has called for the construction of a 15,000 megalitre dam on a tributary of the Mitchell River in East Gippsland. And Victoria's Liberal Party promised in its 2006 election bid to build an Arundel dam (at a cost of $80 million) in a ravine on the Maribyrnong River near Melbourne Airport.

Melbourne Water engineer Geoff Crapper has said in the media that an Arundel dam would provide enough water for 90,000 houses and protect residents in the lower reaches of the Maribyrnong from flooding.

The Bracks Government has ruled out new dams for Melbourne and is sceptical of a regional one. Victoria's Water Minister, John Thwaites, says dams are too expensive to build and cause irreparable damage to river environments.

Dams can't change the fact that water is limited and we have to learn to live with less of it. We have a water shortage because of climate change, growing populations and unsustainable irrigation water use – not a lack of dams.

We need to get better at using rainwater where it falls. I think innovative, small-scale stormwater storage projects are the way forward. Let's invest in retrofitting buildings to use less and harvest more water. We need systems that capture local stormwater, decrease dependence and effects on surrounding catchments and reduce energy use by not having to pump water over large distances. Let's think about that instead of new dams.

OUR LEAKY INFRASTRUCTURE

Some Australian towns lose up to 30 per cent of their water through leaky mains

On the Australia Day long weekend in 2007, a water main in Melbourne burst at 7 am. It couldn't be reported until 9.30 because of a fault in water authority Yarra Valley Water's phone system, so it gushed for hours. Residents who had been diligently conserving water were irate over the waste.

Much of Australia's water infrastructure is more than 100 years old and seriously out of date. It loses gigalitres of water every year in leaks. Our water industry needs tens of millions of dollars of investment, but solutions can certainly be found.

The Harvey Irrigation Area is a great example. It is a farmland region of around 10,000 hectares, south of Perth. A third of its water was being lost in leaky channels before it reached the farm gate. The Harvey farmers knew that Perth was low on water and persuaded the city water utility to pay to replace those open channels with huge pipes, and the water saved was traded to the city. The farmers receive the same amount of water they did before, but now it is available on tap and under pressure.

There is plenty of potential for this kind of win-win investment across Australia – we just need to do more research to understand the impact of leaky systems and what to do about them. John Marsden, a water economist who advises public utility organisations on pricing, property rights and environmental resource governance, had this to say on the ABC's *7.30 Report* early in 2007: 'We know that losses in the big irrigation systems are meant to be around 20 per cent to 30 per cent, but we're not actually quite sure [...] how much of that is measurement error and how much of that is actual losses and, if it's a loss, is it going to the local wetland or is it being wasted and just increasing salinity. We need to know a lot more ...'

Our leaking infrastructure is a big problem, but it can only be sorted out at a state or federal level. In the meantime, though, what can you do as an individual? Well, if you see a leaking water main on the street, call your EPA or water authority immediately. Hopefully you'll be able to get through!

STORMWATER
REUSE

In urban areas, up to 90 per cent of our rainfall flows into the stormwater system

It's hard to watch water rushing down the street and away from us into the ocean when we need it so badly, but stormwater is just not that easy to hold on to. It's unpredictable: sometimes a sprinkling, sometimes a thundering flood. And even when we can catch it, there's the problem of storage.

Land in urban areas is expensive, and there's not much room to build big tanks or reservoirs – or treatment plants. Stormwater runs over contaminated surfaces and then needs to be treated before it can be used.

Canberra and Adelaide have cleared these hurdles by building wetland ponds and urban lakes that collect stormwater and allow suspended matter to settle out. Wetland ponds are very efficient at improving water quality. It helps if they are near aquifers (underground water stores), because the clean stormwater can eventually be injected into the aquifer for storage and pumped out as needed.

A big stormwater project was announced in Adelaide in August, 2006. Thousands of litres of stormwater that would otherwise flow from Adelaide's northern suburbs into Gulf St Vincent in South Australia will be recycled. The scheme will reduce demand on the Murray and provide recycled water for industry and irrigation. The northern suburbs run-off is currently about 62 gigalitres. The recycling project aims to capture about half of that. The water will then be injected into an aquifer about 190 metres below the ground.

The project isn't cheap. It's estimated to cost $90 million, but a number of groups are chipping in. The National Water Commission has provided $38 million. A further $21.7 million will be contributed by local councils, $16.4 million by the state government and $14.1 million by the private sector.

The federal government is funding a number of similar projects across the country under the Urban Stormwater Initiative, but every community is different and no two stormwater reuse plans will be exactly the same. Each community should be assessed for suitable stormwater storage sites before the idea is rejected. Storage solutions can often be found, such as old dams.

Adapting existing urban designs for stormwater harvesting is often costly and difficult, but new communities are being planned with this goal in mind. For example, roof drainage can be directed onto gardens and lawns and run-off from roads and paved areas can be channelled onto grassy areas. New developments are also incorporating holding ponds and detention areas that help filter some of the large debris out of stormwater and promote the settling of sediment.

Q&A *'Does stormwater recycling pose health risks?' asks Ayesha, a cartographer who is always on the move, travelling and mapping Australia.*

Yes! But we can manage them.

Stormwater blasts through our cities and carries high levels of pollution: litter, sewer overflows, vehicle emissions, animal faeces, garden fertilisers, silt and vegetation.

While natural ecosystems can absorb some pollutants, metropolitan centres produce waste streams that are too concentrated and fast-moving to be assimilated by the lakes and oceans they flow into. This can result in algal blooms, the death of marine life, shrinking fisheries and closed beaches. Sensitive marine environments such as Moreton Bay and the Great Barrier Reef are particularly vulnerable to effluent and stormwater pollution.

There are also hidden pollutants in stormwater derived from pharmaceutical products, as well as the chemicals and antibiotics in agricultural run-off.

The good news is that it's now possible to design communities to reintegrate stormwater flows into urban water cycles, allowing this water to be used as a resource. Developments such as the Lynbrook Estate in Victoria use water-sensitive urban design that filters stormwater and creates beautiful wetland and lake spaces.

At Lynbrook, the residential design slows down the run-off and allows more rainwater to be absorbed back into the soil. Drainage water is filtered through grass swales and gravel-filled trenches in the streets and is then treated in wetlands before it enters existing stormwater systems.

In some other towns and estates with similar set-ups, the filtered water is stored in lakes for harvesting or injected into an aquifer.

Australia's stormwater infrastructure was originally built to reduce flooding. Much of it will reach the end of its useful life over the coming twenty years and this provides us with a rare opportunity to replace it with more ecologically sustainable systems that will make it easier to capture and harvest stormwater.

How much is all this going to cost?

Reusing stormwater is expensive. Even the humble rainwater tank has hidden costs: when you buy a tank, you also have to spend money installing and maintaining the systems that will carry the rainwater throughout your house.

As long as we look solely at the cost of stormwater or wastewater reuse versus the cost of drinking water, recycled water will always appear more expensive. That's because drinking water is heavily subsidised. Maybe it's time to start comparing the cost of stormwater reuse to the cost of developing alternative sources of water, and not the current price of drinking water, which many people believe is too low.

Triple-bottom-line analysis gives environmental, social and economic concerns equal weighting. Perhaps this is the approach we need to take when considering the benefits of harvesting stormwater.

Eeeek! It's all too new. Maybe we should wait a few years and see ...

Possibly one of the greatest hurdles with all water recycling is the fact that, to many people, it's 'new-fangled technology'. People will be uncertain about it until it is tried and tested.

David Suzuki loves to talk about his friend, architect Bing Thom. Thom put forth a proposal to the Chinese government to build a community for 1 million people that would be water, waste and electricity neutral. Suzuki said government representatives were fascinated by the idea but first wanted to know where else in the world such a community had been built. Specifically, they wanted to know where in America such a community was operating. Thom told them it was new, it had never been done before. He didn't win the bid.

We must have a go to show that this can be done! Australia already boasts a few trailblazing projects. It would be great to see more architects and planners following their lead. Let's look at a couple of examples.

Figtree Place

Figtree Place, in inner suburban Newcastle, is an amazing example of stormwater reuse in a residential and commercial setting. There are twenty-seven residential units, and the community uses rainwater tanks, infiltration trenches and a central basin in which treated stormwater enters an unconfined aquifer.

During the planning phase of the development, it was decided that water harvested on site should meet 50 per cent of the community's needs for hot water and toilet flushing, and 100 per cent of its domestic irrigation needs. Figtree Place also supplies a neighbouring bus station with 100 per cent of the water it needs to wash its vehicles.

The main features of the development include:

- Underground rainwater tanks, with capacities ranging from 9 to 15 kilolitres, fitted with 'first flush' devices that discard the first part of the inflow, which carries sediment, leaves and other pollutants. Each tank services between four and eight homes.
- Recharge trenches on nineteen of the home sites, each trench 750 millimetres deep by 1000 millimetres wide, and containing gravel 'sausages' enclosed in geofabric. (Geofabric is a tough yet permeable material used for reinforcement and drainage purposes.) These trenches receive overflow from the rainwater tanks and help to recharge groundwater.
- Diversion of the run-off from paved areas to a central basin, which is again used to recharge groundwater.

When Figtree Place was completed in 1998, the project planners hoped to achieve an overall saving in mains water demand of up to 60 per cent. By 2000, this target had already been reached.

Kogarah Town Square

Kogarah Town Square in Sydney's southern suburbs was redeveloped in 2003 as part of Kogarah Council's shift towards sustainable development. The site has 193 residential apartments, 4500 square metres of retail and commercial space, a public building, an underground car park and both public and private gardens.

Water-sensitive urban design concepts were incorporated into the original design, ensuring the capture, recycling and reuse of all stormwater from the site for irrigation, toilet flushing, car washing and a water feature in the town square.

The water passes through a pollutant trap that separates out litter and debris and then flows through a series of tanks. This water is pumped under pressure and used for watering greenery in the large courtyards.

The landscaped areas act as a filter for the water, removing the excess nutrients and fine particles. The filtered water is collected and stored in a separate tank and is used again. This saves more than two megalitres of town water each year that would otherwise be used for irrigation.

The design uses the landscape to filter the water, so that excess nutrients and fine particles are retained by the soil. The 'clean' stormwater (predominantly from roof surfaces) is retained in a storage tank, and passes through a filter and disinfection unit prior to use for higher level needs.

Projects like Figtree Place and Kogarah Town Square are showing the way, and there are loads of similar schemes around the country. Unfortunately, though, water-sensitive urban design principles aren't being employed as often as they could be, even in new developments, where the implementation costs are much lower. What can you do about this? Well, if you're looking into buying a place in a housing estate, check to make sure the developers haven't drained a natural, functioning wetland and replaced it with small ornamental lakes that have no filtering or purifying role in stormwater management. If they have, and you decide against the estate for this reason, let the developers know. Buyers' opinions can influence the design of future projects and make environmental sustainability the new national standard.

Queenslanders will soon be drinking recycled water

It's not just individuals who are taking up the challenge and recycling their greywater. Water is being recycled on a large scale as well, with the federal and state governments, big business, small business and research organisations all getting involved in projects that are extending the life of our water. Recycling water on a large scale has clear benefits. It reduces demand on mains water, and also the volume of effluent being released into the waterways.

Community greywater systems

Throughout Australia, new estates are being designed for maximum water and energy efficency. Many include recycling plants, set up to reuse the greywater of the whole community, sending back for use in toilets, washing machines and so forth.

Rouse Hill

Rouse Hill, a new suburb in the north-west of Sydney, is a great example. A treatment plant was established there in 2001; it was originally intended to service 16,000 homes, reducing the community's demand on mains water. The plant now delivers an average of 1.3 billion litres of water to over 35,000 homes. It is the largest water-recycling scheme in Australia, delivering water to residents' toilets, washing machines and external taps for the garden and outdoor use.

The system delivering mains water and recycled water is what is known as a 'dual reticulation' system, with two sets of pipes. The recycled water is carried by purple-coloured pipes for easy identification. The treatment plant uses a micro-filtration technique that chlorinates the water at a final stage, delivering very high-quality water.

♦

Fourteen large-scale recycling plants around Sydney are currently recycling 15 billion litres of water per year for its residents. The aim is to recycle 70 billion litres of water by 2015.

With its large population, Sydney is leading the way, but similar schemes are happening in Victoria and South Australia, too.

Aquifer recharging

Recycling of water back into aquifers (groundwater wells) is occurring in Perth and other Australian cities.

The Western Australia Water Corporation and CSIRO are conducting a project investigating managed aquifer recharge. Treated wastewater is injected into groundwater aquifers and then reclaimed for further use.

The three-year trial is being carried out at CSIRO's Floreat site in Perth. About 50,000 litres of treated wastewater from a nearby treatment plant are pumped into the managed aquifer each day. The water is tested for pathogens and contaminants during infiltration and storage. The aim is to assess the human and environmental risks involved in storing water in this way for future use. Initial reports from the study are positive.

Recycling water for agricultural use

The Northern Territory's Power and Water Corporation has committed $10.4 million to the innovative Water Reuse in the Alice project.

The project will see an end to treated water from wastewater ponds overflowing into the Ilparpa Swamp during dry weather, and evaporating. Rather than letting precious water go to waste, the effluent will be recycled so it can be reused.

Initially the project will recycle 600 megalitres of water a year and pump it down to the Arid Zone Research Institute, where it will be stored underground before being used to irrigate horticulture projects, helping create employment and economic opportunities for the region. Power and Water are working with the Department of Primary Industry, Fisheries and Mines to complete this important project. Construction is ongoing.

Recycling water to drink

In 2006, the people of Toowoomba famously rejected a major push to recycle their sewage for use in all household fixtures as well as for drinking. Around 60 per cent of residents in the inland Queensland city voted against the proposal. But only six months later, on 28 January 2007, Premier Peter Beattie announced that Queenslanders are going to have to get used to the idea of drinking recycled water. The premier reneged on a $10 million plan to host a referendum involving eighteen councils to gauge whether residents would accept drinking recycled water. Beattie said that record-low inflows to dams had left him with no choice but to introduce recycled water into the state's drinking supplies. Recycled drinking water will reach residents by the end of 2008, after all of the necessary infrastructure has been put in place.

The reality is that residents in many major cities such as Singapore, London and Washington have been drinking recycled sewage for some years. Beattie says all of Australia is going to end up drinking recycled water. Hopefully his actions will inspire other states to follow – and sooner rather than later.

'These are ugly decisions, but you either drink or you die. There's no choice. It's liquid gold; it's a matter of life and death,' Beattie told ABC radio.

Even though the federal water minister and the Howard Government came straight out and supported Beattie's commitment to recycled water, there has been some opposition by experts who say it isn't safe to drink treated sewage. The main fear is that technology may fail due to human error.

South Australia's premier, Mike Rann, is against the idea for his state. On ABC regional radio, he said, '… the plan is to allow the cotton growers and rice growers upstream to use pristine river water, but at the same time saying that the treated sewage effluent water should be used for drinking water in our capital cities, well, I can veto that in South Australia.'

Adelaide already recycles over 20 per cent of its effluent. Some of the major uses are on the northern Adelaide Plain and on the vineyards near Willunga.

Victorian Premier Steve Bracks says the use of recycled water is essential for irrigation in his state, but not for drinking.

David Campbell, who was then the New South Wales Minister for Water Utilities, told the media that there's no need for Sydney's families to drink recycled sewage, as the government is able to secure the city's water supply. He said it is up to local councils in rural regions to decide for themselves, based on their local circumstances, as they have the authority over their own supply.

So it seems Australians are still reluctant to drink recycled water. They have voted against it where they have had the choice, even though many countries are safely doing it already. We need a concerted education campaign. After all, anyone who has ever been to London has drunk recycled water – and that's a hell of a lot of us!

Recycling versus desalination

Recycling and desalination both have advantages and disadvantages. (We'll look at desalination in the next chapter.) Sydney Water estimated last year that it would be approximately 70 per cent cheaper to build a desalination plant than to build a recycling system. The main problem with recycling was that the treated water would have to be pumped almost 60 kilometres uphill to the Warragamba dam for storage. However, there was concern that operating a desalination plant could be extremely expensive: the costs of running a recycling program would only be about 60 per cent of the costs incurred in running a desalination plant.

If we all do our bit on a domestic level and if state governments continue to build large-scale recycling plants, we won't have to build dams, or big desalination plants that will be costly to run. I believe that recycling is a more logical step than any other of the major 'band-aid' solutions. All of us can recycle and we can do it at a relatively low cost.

DESALINATION: THE TECHNOSOLUTION?

We can have as much water as we are prepared to pay for

When I get everyone at the bar talking about the water crisis, there's always a handful of people who say, 'Don't worry about it, mate, we're going to be drinking seawater before we run into trouble. We've got oceans full of water – this is not a crisis!'

Desalination seems to be everybody's favourite solution – if only it were that easy!

What is desalination?

Desalination is the filtering of salt water into safe, clean drinking water. Like recycling, it is a drought-contingency measure that's not dependent on rainfall.

Desalting seawater is not a new idea. Aristotle described an evaporation method used by Greek sailors of the fourth century BC. In the nineteenth century the development of steam navigation created a demand for non-corroding water for boilers; and the first patent for a desalination process was granted in England back in 1869. World War II saw further major developments in desalination technology as various military establishments in arid areas required water to supply their troops.

Desalination has changed the way people live their lives and where they choose to live. The change is most apparent in parts of the arid Middle East, North Africa and some of the islands of the Caribbean, where the lack of fresh water previously limited development. The United States, China and Spain are other countries desalinating water and doing it well.

There are now more than 7500 plants in eighteen countries producing approximately 35,000 million litres of fresh water per day. The global desalination market is currently worth $2.6 billion per year, but it's expected to grow to $91 billion per year over the next twenty years.

Is desalination the way to go in Australia?

Amy spoke to Greg Leslie, an expert on desalination from the University of New South Wales, to find out more. Leslie says Australians cannot assume desalination is an easy way to deal with the water crisis. We shouldn't be waiting for a technosolution to the drought. Instead we should look at our daily lives and do something now to reduce the amount of water we use in our homes.

'The World Health Organization says people need 50 litres of water per day as a basic requirement for preparing food, washing and drinking. In Sydney's affluent neighbourhoods people are using up to 400 litres of water a day. In water-conscious homes, people who are really saving water and working hard still use about 200 to 250 litres of water per day.'

Leslie explains why desalination isn't necessarily the ideal solution: 'It comes down to what we call yield. For every dollar we spend you want a certain yield of water. Take Sydney's water desalination proposal [announced by Bob Carr back in 2005]. It was to provide about 125 megalitres of water a day for a price tag of about $1.5 billion. That would provide 10 per cent of Sydney's water supply. If Sydney took that $1.5 billion and said, okay, we're going to upgrade every washing machine and dishwasher, and make sure they are low flow and AAA-rated, they could save more than 10 per cent of the daily water supply and have a lot of money left over.

'A low-flow showerhead, a good one, will cost you about $100. We've got about 1 million domiciles in Sydney, so to upgrade all the showerheads it would cost $100 million. That would save 50 per cent of the water used in showers.

'But Sydney Water says "No, we can't do that, it's draconian." Well, it's time to be draconian.'

The original proposal made in 2005 was shelved, but it was resurrected by the Iemma Government when dam levels fell to 34 per cent over the course of the following year, perhaps because Perth had managed to open a desalination plant without serious protest from the community. In November 2006, plans were announced for a plant at Kurnell in Sydney's south. The plant will use reverse osmosis technology to remove salts and other impurities from seawater to produce drinking water.

Water from the desalination plant will be pumped into Sydney's water distribution system through a pipeline across Botany Bay to Kyeemagh, eventually connecting to the main city water tunnel at inner-city Erskineville. From there the water will be distributed to up to 1.5 million people south of Sydney Harbour.

Leslie says desalination will top up supplies and help the city limp along to the next rainfall, but it's not a sustainable solution.

'It would be a disaster if Sydney started to look like Dubai and Saudi Arabia, where 90 per cent of the water comes out of the ocean and is desalinated. We humans can treat water to any standard we like. We can make any water super-clean. But this method is just too expensive, it takes way too much energy, and we can't rely on it.'

He admits, though, that in some situations, desalination may be the best option available. For instance, he says that Perth got hit 'way too hard with climate change' and there wasn't time to convince people to conserve in their homes. Perth was running out of water and the state decided it had to turn to desalination. Fortunately the city is also devising ways of recycling the water it desalinates. As Leslie says, 'It doesn't make sense to use a lot of energy to desalt water, use it once, and throw it away.'

The price tag

Most desalination technologies consume an outrageous amount of energy, resulting in high financial and environmental costs. Desalinated water is more expensive than mains water as a result, making it an unattractive option for all but the most remote rural water users or those in extreme drought.

It is only cost effective in limited circumstances in Australia:

- where there's no fresh water nearby and it has to be piped in
- where low-cost energy supplies are readily available.

This may change. Many remote towns and communities rely on pricey and often limited supplies of diesel fuel for their energy needs. These and other forms of fossil fuels are sometimes heavily subsidised by governments to meet their obligations to these communities. Now that we're finding ways to use renewable energy to power the desalination process, it is becoming more viable.

As fresh water becomes scarcer, desalination is expected to become more popular in Australia. As water markets develop, the price of fresh water is likely to go up to reflect its true value, making desalination a reasonable option.

Pros and cons

Desalination has some distinct drawbacks. Desalination plants consume a lot of power, and generate noise. They also produce a highly saline brine stream which poses environmental problems, especially if it is disposed of in an estuary or bay, where it can have an impact on marine life. However, communities around the world are now figuring out ways to make money from desalination plant waste.

The waste is actually an asset that can be exploited. Leftover brine water is used in a number of economic ventures, reducing the overall cost of providing fresh water. Saline water can be used to irrigate particular horticultural crops. Near Quorn in South Australia, crops such as olives, almonds and pistachios have been produced for over twenty years while under irrigation with saline. Markets also exist for a variety of aquaculture products that can be grown in saline water, such as brine shrimp and algae. (Beta carotene, popular as a dietary supplement, can be obtained from the algae. Australia is one of the few countries in the world where these algae can be successfully cultivated.)

Salts and other minerals can be extracted from brine by mechanical means or crystallisation, then sold to agricultural, industrial and domestic users.

Finally, salt water can be used to collect and store energy from the sun in 'solar ponds'. The hot saline water can be used to provide consumers with heat or electricity.

Current desalination projects in Australia

Commercial-scale desalination started in Australia in the late 1960s and 1970s. The cost of building and running the plants fell in the 1980s and 1990s as the technology improved and people became more experienced in operating the plants. Nevertheless, we have comparatively limited expertise in desalination technologies and have been slow to implement them.

The main users of desalination in Australia are isolated mining towns, small communities and industry. Because it's so expensive, only a limited number of small desalination plants are used to produce public water supplies, but desalination plants can supply water to populations ranging in size from individual households right up to cities.

I'd like to look at a couple of examples in detail. Perth has invested heavily in desalination, and the New South Wales Central Coast is also making use of desalinated water, but their approaches are very different.

Perth

The challenge faced in Western Australia is not only a changing climate and reduced rainfall but a growing population and a booming economy that creates ever-increasing demand. The WA Water Corporation responded by joining forces with construction company Multiplex and water-treatment specialists Degrémont to build the Perth Seawater Reverse Osmosis Desalination Plant or 'SWRO' at Kwinana.

The Perth SWRO is the biggest seawater desalination plant in the Southern Hemisphere, producing 45 gigalitres of water per year or 130 million litres per day. It provides 17 per cent of Perth's water needs. Most importantly, the desalination plant produces a secure supply of water that does not depend on rainfall. The plant cost $387 million to build. Annual operating costs are less than $20 million a year, and the cost to each household is estimated to be about $44 per year. To reduce its environmental impact, electricity for the plant is being produced on a wind farm located 30 kilometres east of Cervantes.

When the plant opened in November 2006, Western Australia became the first state in the country to use desalination as a major public water source. The SWRO operates by converting half its seawater intake into drinking water. The salt remains in the other half of the water, which is diluted about forty-five-fold as it is returned to the ocean. This salt water concentrate is jetted out under high pressure and mixes rapidly with the surrounding waters.

Studies done by experts at the University of New South Wales during the planning stage showed that the operation would boost the salinity in Cockburn Sound by less than one per cent and that there would be no adverse ecological effect. The volume of seawater carried by the tides flooding in and out of Cockburn Sound every day is 400 times the amount taken out by the desalination plant. In other words, the amount taken out by the desalination plant will be 1/400th or 0.25 per cent of the amount of water moving in and out of the sound on a daily basis.

Sensors have been placed on the bed of Cockburn Sound to enable the WA Water Corporation to monitor the temperature and salinity levels of the water. The Perth SWRO has satisfied all environmental approval requirements so far.

New South Wales Central Coast

The water shortage on the Central Coast is so dire that two councils have been forced to adopt the radical idea of leasing mini-desalination plants. Wyong and Gosford share the same water system and have agreed to install mini-desalination units they can use as a last resort if dams dry up.

The mobile desalination units can supply an extra 8 million litres of water a day, enough to service about 20,000 homes. They're about the size of a semitrailer and can be installed in a few months and easily decommissioned. However, they produce water at about $4 for every 1000 litres, compared with $1.12 now charged by the joint water authority. The mobile units could cost as much as $90,000 a month to lease. (The price tag for building a permanent and much larger desalination plant would be about $60 million.)

The plan is controversial, but Ken Grantham, a Wyong Council Shire Services representative, says that desalination really is the last resort: the two councils have already banned the use of all town water outside the home. They are tapping into bore water, taking water from the Hunter region and investigating major water-recycling schemes.

With dam levels down to 16 per cent and creeks running dry, the plan won backing from a special committee chaired by the New South Wales Premier's Department. State staffers met Gosford and Wyong councils to help them comply with planning, environment and health regulations. The councils selected beach sites in the spring of 2006.

In Wyong, three temporary desalination sites have been approved, but the machines have not yet been installed. If dam levels go below 10 per cent, the council will be able to act right away and get the units up and running because they already have the approval to go ahead.

In Gosford, eight potential sites for temporary desalination units are still being investigated. The development applications for these sites have not yet been approved.

The councils have since decided they need to find sites for more desalination units to supply another 25,000 homes.

The plants will have some effect on the environment, and possibly on the area's tourism industry. The energy a single unit would require to produce 2 million litres of desalinated water a day, servicing 5000 homes, would create the same volume of greenhouse gas emissions as each of the 5000 households driving a small four-cylinder car about 7 kilometres a day, according to the two councils.

The slow rate at which the seawater is extracted from the ocean is expected to reduce the likelihood of fish, larvae and other marine life being drawn into the beach wells. A number of noise control measures have been included in the overall installation process to minimise the impact on the surrounding neighbourhood.

When the units on the beach have been set up, their effect on marine life, their power consumption and noise emission will all be monitored.

Proposals for other Australian cities

As mentioned above, a desalination plant is on the cards for Sydney. In February 2006, the New South Wales government issued a request for tender documents for the construction of a 125 megalitre per day desalination plant at Kurnell. The desalination plant will be powered entirely by green energy, and, according to Sydney Water, the construction methods and site will be designed to minimise impact on the coastal environment, water quality and the seagrass beds.

The Queensland government is working with the Gold Coast City Council to build a desalination plant at Tugun by the end of November 2008. An early works program is already underway, and it's estimated that the final costs will come in under $1.126 billion. Ongoing operations and maintenance will cost up to $40.8 million per annum, averaged over ten years, if the plant is run at full capacity for the entire year.

ACTEW, the energy and water provider for the ACT, has already completed a preliminary exploration of the potential for building a recycling plant that would allow the ACT to desalinate and bring up to drinking quality a proportion of the water that is now released into the river at the Lower Molonglo plant. With dam inflows in 2006 almost 90 per cent below the expected average, ACTEW has had to explore moving the recycling plans forward some ten years earlier than expected. The ACT will make a request for assistance from the $2 billion Australian Government Water Fund, should it choose to proceed with the proposal.

Melbourne Water is currently working on a feasibility study on a desalination plant to provide an extra 300 million litres of fresh water a day, at a cost of around $1 billion or more, but as Melbourne is situated on a bay, not the ocean, it seems likely that desalination will prove an expensive solution. If a plant is sited on the ocean, costly infrastructure will be needed to deliver the desalinated water to the main water supply network.

The other big Australian cities are unlikely to give desalination serious consideration in the near future. Darwin and Hobart have no need to supplement existing catchments, and Adelaide's water resources should be sufficient until at least 2050, even if its population doubles.

◆

Desalination might initially seem like a quick, easy fix, but we need to weigh up all the pros and cons of the technology before making a commitment to it. As Malcolm Turnbull, the Federal Minister for the Environment and Water Resources has said:

> The bottom line for our cities is simply this: we can have as much water as we are prepared to pay for. There are many options to secure that water and we need to be careful to make our water decisions rationally, both in an economic and environmental sense.

AGRICULTURE: THE GREAT DEBATE

*More than half of Australia's
130,000 farms are now within
drought-declared areas*

I have friends who scour the fine print on bags of rice at the grocery store looking to see where it's made. If it's Aussie rice, they won't buy it. People see rice growing in fields flooded with water and assume it's wasteful and shouldn't be grown here. Too many urban and suburban Australians point the finger at agriculturalists, blaming them for the water crisis, because they don't know the facts.

People who live in the city need to make a bigger effort to understand the effect the drought is having on rural communities. They are really doing it tough. I know a woman who wears a button on her jacket that says: 'Farmers feed cities, but can't afford to feed themselves.' It's true: farmers fall victim to the drought first, before the rest of us. We should try to remember that before we complain about restrictions that stop us watering our lawns.

It's hard to get the facts on the use of water in agriculture, but that's understandable, because it's an issue subject to a lot of spin. Farmers and environmentalists tell very different stories about how

A view from the country

Yvonne is a rice farmer in rural Victoria, and lives in a community that has been hit hard by the drought.

Yvonne: I wish people could see the big picture. There's a lot of hardship and suffering out there because of the drought. It's not only the farmers who are struggling. It's the processing companies, machinery businesses ... It's really hit the whole community.

Farming communities are pretty close-knit. We've been holding our own, but morale has been low. Generations of hard work has been lost to drought. There's a bit of desperation about where things are going and what's going to happen. The social impact on communities is quite extensive. Marriages break down, wives and kids leave the family farm to find jobs. People get suicidal.

Many of our Australian farmers' kids haven't actually seen rain. There are some areas where it literally hasn't rained for years. For nearly a decade, families in rural and regional areas have been digging themselves into a deeper hole, financially and emotionally. It's been going on for a long time now, and sometimes it feels like the cities have just forgotten about it.

When we do have rain, it lifts people's spirits. A large downpour brings back hope that things are going to change. But farming is cyclical. If you think, okay, we've had some rain, I think the weather's going to change, and you plant some wheat, you're not going to see a return on that crop for five months. And by that time we may not have had another raindrop and all that wheat could have burned off. It's the same with the fruit and veg industry. Even if our farmers get rain now, there are people who are going to be in hardship for years to come.

much water is used to grow controversial crops like rice and cotton in Australia. At a Melbourne University forum about water, a representative of the Victorian Farmers Federation said it took 200 litres of water to produce a kilogram of rice. A water spokesperson from the private sector argued that the amount was closer to 16,000 litres. It's difficult to know who to believe!

The savewater! Alliance is a non-profit association working with water businesses, government agencies and manufacturers to deliver water conservation programs throughout Australia. They are widely considered the highest authority on water in the country, so I'm going to go with their figure. They note that the maximum water allowance in the Murray region is 4 megalitres per hectare and the average yield in the area is around 9000 kilograms per hectare, so they calculate that an accurate figure would be about 450 litres per kilogram of rice.

But it doesn't matter whether the answer is 200, 450 or 16,000 litres. It's no use saying outright that rice and cotton should not be grown in this country because they require too much water. The rice and cotton industries combined bring in around $2 billion in export revenue per year. Most of us know that the environment is more important in the long run than money, but it's also important to recognise that agriculture provides us with the food and clothing that we need to survive.

Water use in agriculture

It's true that cotton and rice crops use a lot of water, but it helps to know the total breakdown across the entire agricultural sector. According to the savewater! Alliance, the biggest water users are livestock, pasture and grains producers. Together they account for 56 per cent of the sector's total water use. Cotton is next, at 12 per cent, followed by rice at 11 per cent and sugar at 8 per cent. Fruit accounts for 5 per cent; that figure excludes the water used on the grape crop, which accounts for 4 per cent of the total. Vegetable growers use the remaining 4 per cent. This is illustrated in the graph below.

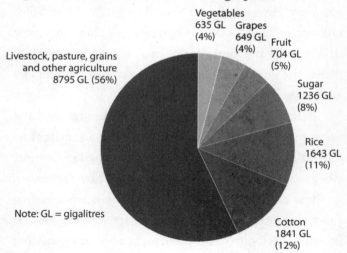

Net water use in the agricultural sector.
(Source: savewater.com.au)

The agriculture industry has put a lot into saving water in recent years and deserves recognition for this. There have been rapid improvements. For example, rice production has increased by 60 per cent over the last decade, while water use has decreased by about 30 per cent.

I would like to touch on a few of our agricultural industries that have been accused of wasting water – sometimes for good reason – and tell you how they are doing now with less H_2O.

Rice

Rice farmers make an important contribution to our economy. Annually, their crop brings in about $800 million, so it's difficult to say we simply shouldn't grow it. About 85 per cent of rice in Australia is exported, while the average Australian consumes around 10 kilograms of rice per year.

In the past, the industry had a reputation as a water guzzler, but it has been put under a spotlight as a result, and forced to cut down. Over the last decade, rice growers have slashed their water use by a third.

It is possible to use water more efficiently when producing rice and farmers who do are enjoying several benefits: they don't run their pumps as often, cutting down on energy consumption; they have fewer problems with salination, which often results from overwatering;

and if they are able to sell their excess water on to others, or employ it elsewhere, they achieve greater profits.

Rice farmers can and are controlling water use. Many now stop watering just prior to the full maturing of the plant so that every last drop of water is soaked up. Some farmers are also trying off-season crops, such as wheat, to take advantage of the remaining moisture in the soil after the rice harvest.

In the last few years 95 per cent of farms have started overhauling their operations in order to save water. Half have already completed their changes. The industry is working as hard as any other in its efforts to conserve water, and Australian rice is now the most productive per megalitre of water in the world. This doesn't mean that the industry has reached its water-saving potential, though. There is still a lot more to be done.

Cotton

Cotton, like rice, has been tagged as a major baddie when it comes to water use, with some justification, but it's hard to ignore the fact that the cotton crop pulls in around $1.8 billion per year in export revenue. It's the fourth-largest rural export industry in the country – about 94 per cent of Australian cotton is sent overseas.

Australians are actually the greatest cotton users per capita in the world, largely because cotton can be spun into comfy, breathable clothes that work well in our hot weather. We sell cotton in bulk to other countries where it is sewn together into garments and sold back to us.

There are about 800 growers covering a total of 536,000 hectares of farming area, but due to the drought only 220,000 hectares have been farmed in recent seasons. Cotton growers have been seriously streamlining their water practices: the average yield per hectare in Australia is now 1600 kilograms per hectare, approximately 2.6 times the world average.

Dairy

The dairy industry produces 10,090 million litres of milk per year. The industry has a lot of pull because it raises around $3 billion a year in export revenue. It uses around 20 per cent of all allocated water supplies in the primary industries.

There are two different types of dairy farm: dry and irrigated. Many dairy farms are dry farms and rely only on the water that falls from the sky, but irrigated farms are a major user of local groundwater supplies. Almost all the water used on dairy farms goes toward irrigation to grow food for the cows.

While researching this book, I spoke to dairy farmer Arthur Munro. He runs a dry dairy farm in western Victoria that accommodates up to 200 dairy cows at

Arthur Munro runs a dry dairy farm in Victoria's western districts. The dairy has been farmed for 106 years, and Munro has produced milk there since 1993. Historically, the area has enjoyed seasonal rains, but Munro now has to pump water from a nearby spring to keep enough water in the dams. The cows drink the water, and it's also used in the shed for cleaning.

Every year Munro has been on the farm, the drought has worsened, forcing him to pump more and more water from the spring. The pump he uses consumes about 130 litres of diesel every day.

When I spoke to him, he explained the difficult situation he is in.

Arthur Munro: The evaporation lately has been the worst we have ever seen. On a hot day the air is so dry that you can almost watch the dam level fall. Because of the drought we've had to stop milking this year in early January, instead of stopping some time in February. We still have to feed the cows, and that costs. But for that month or six weeks we have no money coming in, because the milking is finished.

one time, slightly larger than average farm in the area. Each cow drinks around 80 litres of water per day, and operating and maintaining the shed uses up around 3000 litres per day. Altogether, Munro needs a total of around 19,000 litres of water per day just to run the farm.

During periods of drought, the farmers in the area are forced to pump water from a nearby spring. Munro and his neighbours also face another problem: big woodchipping companies have bought out many nearby properties and are planting blue gums. The gums drink up a lot of water from natural and local watercourses, reducing the total amount available to the farmers.

Dairies have had to cut back on water use, just like other farms. They haven't really had a choice!

Viticulture

The wine industry earned around $1.75 billion in 2001 and aims to increase annual earnings to around $4.5 billion annually by 2025. The big dry has actually had an upside for winemakers, because it has reduced the wine glut. Bumper crops in recent years left many wineries and retailers with an excess supply, prompting them to slash prices. Water restrictions and frosts in the southern grape-growing regions have eaten into production and helped to reduce stockpiles.

Nevertheless, winemakers stand to benefit from saving water just like rice and cotton farmers. Efficient irrigation will help them increase crop yields while freeing up excess water to expand production.

In a lot of wineries, meters are being installed to monitor water use and detect possible leaks. Flow restrictors are now being used to control flow rates on all equipment and taps, and many farmers are treating water that has been used in the winemaking process and reusing it for irrigation.

Horticulture

Horticulture is one of Australia's fastest growing industries. The production of fruit and nuts, vegetables and nursery products brings a huge amount of money into the Australian economy – about $6.7 billion per annum.

According to the Horticulture Water Initiative, for every 100 megalitres of water the industry uses, it generates around $250,000, creates four jobs and adds $1 million to the economy. To continue to grow at its current rate as water supplies dry up, the industry will have to employ more efficient irrigation techniques, but this is already starting to happen. Drip irrigation has reduced water consumption and many horticulturalists are using recycled water on their crops now.

Livestock, pasture and grain

The graph on page 287 showed us that more than 50 per cent of water used in agriculture goes toward livestock, pasture and grain-related industries. Half of that – or 25 per cent of the agricultural sector's total water use – is used to produce dairy products and meat.

Consider how much land had to be cleared so that it could be used to graze livestock. Then think about the amount of greenhouse gas produced by the animals themselves. Add that to the growing cost of transport as fuel becomes more expensive, and it would be fair to ask if the Australian meat industry is truly sustainable – though arguing that in public may leave you open to ridicule, never to be taken seriously again! We argue and argue about rice and cotton … what about dairy and beef? A case for vegetarianism?

It takes 22,000 litres of water to produce about half a kilo of meat. By not eating half a kilo of meat you could save almost as much water as you would use in a shower in twelve months. The average vegetarian diet requires the use of about 1320 litres of water per day, while the average diet of a meat eater requires around 17,000 litres of water per day. (John Robbins' *Diet for a New America* is a great book on this topic.)

Nevertheless, we have to remember that the Australian meat industry has an annual export value of around $7 billion, which accounts for around 30 per cent of our total annual food export. The livestock sector is a huge water user, but it creates thousands of jobs that keep our economy moving forward and supports many Aussie families. Giving up meat might be hard!

Looking to the future

Despite our farmers' best efforts, there is little doubt that we need to continue looking at alternative ways to manage our resources – no matter how radical they may seem.

Tasmania and the Northern Territory are two states that get pelted with very high levels of rainfall at particular times of the year. I had a hard time acquiring information on water conservation in these areas of the country while I was researching this book, and I wasn't surprised. Rainfall has actually *increased* in some parts of Tassie and the Northern Territory as a result of climate change. That presents us with options as we consider the future of agriculture. The federal government has suggested that if we start to build infrastructure in the Northern Territory now, the great vast land could become the new centre for farming in Australia.

Liberal senator and farmer Bill Heffernan recently said, 'We have to create a new agricultural frontier …

a vision similar to the Snowy.' Senator Heffernan says that if we could harness just 10 per cent of the water that falls in the Gulf of Carpentaria and Timor area, we'd have more water for farming than we do in the Murray–Darling Basin. He'd like to see a new generation of farmers move north. They would need incentives, but large tax breaks plus the opportunity to make a decent wage should do the job.

It's a great idea. But we must learn from the mistakes that we have made in the Murray region in past years and be very aware of the environmental risks.

We should remember, too, that moving our nation's food bowl to the Northern Territory is not merely a question of environment and rainfall. Much of the land in question is owned by Indigenous Australians, who have had to fight hard to maintain their rights to it. We can't just bowl into any area we see fit. We have created our problems and we need to find solutions to them. As a society, we could learn a lot about land and resource management from our country's traditional owners.

In January 2007, bestselling author and world respected scientist Tim Flannery spoke with Kerry O'Brien on the *7.30 Report*, and asked why Australians don't place more importance on country. Flannery said he was puzzled by the way we bang on about 'meat pies and football and Holden cars' when our land is the one thing we all share – 'what gives us our water and our food and our shelter and defines us as a nation'.

I think the answer lies in our nation's past: when non-Indigenous Australians' ancestors came from overcrowded England and elsewhere to this vast country, they thought the land and resources would last forever. As Flannery has said, 'It was as if we ate through the wealth of the country in just a few decades rather than carefully shepherded it.'

We are only now starting to see the effect that we have had on our environment, coming to terms with the sensitivity of our ecosystem, and beginning to look at it in a new light. About time! It's essential that we learn from the mistakes of our past if we are to survive the current water crisis.

RISKY BUSINESS

A typical 300-room hotel uses up to 225,000 litres each day – that's 1.3 Olympic swimming pools or 750 litres per room!

Industry has a bad rep when it comes to the environment. People seem to love calling it 'dirty' and many assume industry consumes, pollutes and has no social conscience. But most of us don't know who's using what when it comes to water. I've found it's best not to criticise until you know the facts, so here's a run-down.

Industry uses only a fraction of Australia's water, but the CSIRO has predicted that this fraction will go up. Manufacturing consumes 3 per cent of the country's H_2O, mining uses 2 per cent and the generation of electricity and gas uses 1 per cent. Agriculture sucks up the vast majority of Australia's water supply: 65 per cent. Households use 11 per cent.

Australia-wide, water use fell over the five years between 2000 and 2005, the drop largely attributable to savings in the agricultural sector.

I looked at agriculture in the previous chapter. Here, we're going to look at a few other industries.

While Australian households cut their water consumption by around 8 per cent between 2000 and 2005, industries like mining and manufacturing used more. In fact, mining used approximately one-third more water in 2004/05 than it did in 2000/01. This is because there has been more mining activity, particularly in Western Australia, where there was an 81 per cent increase in total water use by the mining industry between 2000/01 and 2004/05.

Industry	Gigalitres consumed 2000–2001	Gigalitres consumed 2004–2005	Increase/ decrease
Agriculture	14,989	12,191	↓
Electricity and gas	255	271	↑
Forestry and fishing	44	51	↑
Manufacturing	549	589	↑
Mining	321	413	↑
Water supply	2165	2083	↓
Other industries	1102	1059	↓
Household	2278	2108	↓
TOTAL	**21,703**	**18,767**	↓

Source: Australian Bureau of Statistics: 4610.0 – Water Account, Australia, 2004–05

There was also a 9 per cent increase in the manufacturing industry over the same period, with the largest increase in metal product manufacturing. However, the biggest users in the manufacturing sector are the food, beverage and tobacco companies. The makers of metal products come next, followed by wood and paper products and petroleum, coal, chemical and associated products.

Electricity generators also use a great deal of our H_2O. Most of it goes towards hydro-electric power generation. Water used to produce hydro-electricity passes through turbines to generate energy and is then discharged and made available to downstream users. (The scientists and statisticians monitoring water use acknowledge this reuse when collecting data: water used for hydro-electric power generation is treated differently from other water uses and is called 'in-stream' use.) However, coal-fired power stations also use a lot of water in their boilers and cooling towers. Evaporation from cooling towers is responsible for much of the water consumed in the production of electricity.

Who are our biggest water users?

Governments are now trying to drag the big water users out of the closet. The Victorian state government passed legislation in 2006 which would allow the public identification of the state's top 200 water users. The plan was that large industrial water users would be identified, but they would also be given a chance to demonstrate their water-saving efforts to the wider community. A company that uses a lot of water isn't necessarily inefficient and wasteful, of course, but it's still fair to ask large users how they are contributing to Australia's overall water-saving effort.

However, in January 2007, the state government said it would be 'inappropriate' to name the companies on the Top 200 Biggest Water Users list. Instead, the state is calling on companies on the list to come out voluntarily and tell the public how much water they are using. The government's retail water authorities City West Water, South East Water and Yarra Valley Water launched 'Pathways to Sustainability', a program encouraging Melbourne's top 200 water users to prepare water management plans which identify ways they can conserve water. The program has already achieved significant savings.

Not surprisingly, the government has come under pressure for keeping the list secret and is now saying it will be unveiled at the end of 2007, when the water authorities table their annual reports. Sustainability Victoria is calling on the government to further amend the legislation to allow for the identification of the top 1000 water users in Melbourne. The group also suggests the program be expanded to include regional businesses.

Victoria's Pathways to Sustainability program has now been expanded to target all users of more than 10 megalitres, around 1500 companies, with cumulative savings projected to reach an extra 13 gigalitres by 2015. The first water management plans will be completed within three years. The program will then be expanded to other significant water users across metropolitan Melbourne.

In New South Wales, the government released the Metropolitan Water Plan for Sydney in 2006. It spawned a program called 'Every Drop Counts' that targets businesses. It now has more than 300 members, who claim to be saving 24 million litres of water a day between them. Sydney Water has recently embarked on a water recycling project with BlueScope Steel in Port Kembla which will save 7.3 billion litres per year.

A roadmap for the future?

We've made it clear that industrial water use accounts for just a fraction of the total amount of water used in Australia, but it's a fraction that has been growing quickly. By 2050 CSIRO estimates that demand for urban and industrial water will almost double. The deal is this: we're facing an increase in demand, and a decline in resources due to climate change.

It's true that industry uses only a small amount of total water supply available – nothing compared to irrigation. But unlike farmers, who get support from urban Australians who like to eat the food they produce, industry does not inspire public sympathy. Industry is often seen as the bad guy, with governments, the media and the community only too ready to point the finger. Smart businesses will see this as an opportunity for industry to show the way to sustainable water management in Australia.

Study after study has shown that there aren't any major technological hurdles preventing our industries from providing that lead. This has been confirmed by the Barton Group, an alliance of CEOs formed to oversee the development of environmental action in Australian industry, who launched a 'Water Industry Roadmap' in Canberra in 2005.

Their roadmap states that while ongoing scientific research is a constant requirement, 'there are benchmark technologies and systems available for adaptation to effectively solve our basic water security problems within a reasonable timescale of a decade or so'. The major barriers to early success, according to the Water Industry Roadmap, are institutional and cultural failure.

Our society may be inspired to conserve when the price of water increases, but the market signals are probably not yet clear enough for businesses to justify the significant investment needed to achieve big changes in how they manage water.

Governments are basically asking industry to take

a leap of faith. We know the direction our climate is headed (straight towards dry, dry and drier), and we know the technological options we have. That just leaves the timing. What if the drought breaks tomorrow? We don't think it will, but anything's possible. The companies that sink a lot of money into water-conservation technology might not come out the winners if that happens; however, companies that take the risk and make the investment now may be able to absorb and overcome the cost layout by the time a severe water shortage hits. They would then be in an amazing position to compete, whereas companies that ignored the warnings could be suddenly hit with huge costs at a time when competition and survival are crucial.

If Australian industry can get this right, it has an opportunity not only to lead the way in Australia, but to export its knowledge and technologies worldwide.

Big business does its bit

Some of the members of the so-called dirty world of industry are doing good things, and not just in terms of saving water. Some are also donating money to people affected by the drought. But they haven't exactly gotten a pat on the back for the effort.

You know the curse of the tall poppy? Australians have a tendency to cut down those who are superior to them. That's all I could think of when Woolworths raised money for drought-affected families and was criticised for it. We should be commending companies that want to help if we hope to inspire others to do the same!

On Tuesday, 23 January 2007, Woolworths/Safeway donated 100 per cent of the day's profits – $4.7 million – to farming families struggling because of the drought. Finally, a campaign organised by a corporation with deep pockets that would not only help farmers but also get city people involved and offer them an easy, no-brainer way of donating.

People complained, saying the contribution wasn't big enough. 'Woolworths didn't advertise the fundraiser well enough.' 'It shouldn't have been held on a low-traffic day like Tuesday.' 'Why didn't they do it on a Saturday when everyone shops?'

On the day of the fundraiser, Amy and I interviewed people at a Safeway store in central Melbourne over lunchtime, asking them what they thought of the campaign. A couple of women working at a nearby Cancer Council store popped in on their break.

'We were just talking about it on the way over here. I'm not sure how I feel about fundraising campaigns where people are asked to purchase products. I mean, Safeway can simply make a donation. Their profit is around $3 million a day. They should be donating every day!' said Marita Black. Still, she supported the grocery giant in its effort by going grocery shopping and so did thousands of Australians across the country. 'I think it's hard to find a concrete way to help when the problem is a drought and it's just so big,' she said.

Woolworths acknowledged that it wasn't an expert on farming communities and their needs and transferred all the money it raised to the Country Women's Association for distribution to struggling farming families. One third of the money will go towards research into sustainable farming practices.

Families will be able to use the money to pay utility and doctor's bills, to buy groceries and school supplies for their children. The sums distributed would probably have helped a family keep going only for about a month, but CWA president Lesley Young was excited.

'Australian people rally together in times of adversity and this amazing contribution is evidence that we can really pull together when we need to.

'We have seen the best of Australia and we have been reminded that there is a strong and enduring link between the country and the cities. All Australians are suffering this drought together,' Young said, adding that the initiative raised the spirits of people in the country 'beyond belief'.

Young hopes the campaign will mark the beginning of a new era of agriculture in Australia, one in which businesses get involved, fostering communication and understanding between and farmers, corporations and consumers.

She said she was disgusted by emails she received the day of the fundraiser criticising Safeway for not doing more and simply deleted them. A lot of customers shared her view.

'I'm not going to be cynical about it. I think they're trying to do the right thing and that's fantastic,' said Julie Sellenberg, as she carried bulging green bags full of groceries to her car. 'We should support and recognise companies when they do the right thing.' She said she'd stocked up on as much as she could.

Others simply bought what they could afford for the day and went home happy, knowing their dollar was going towards something good.

'Everyone is feeling pinched these days and it's hard to find ways to give,' commented Sally Morris, a young woman shopping with her little boy. 'I think what Safeway is doing is amazing. It gives us all an opportunity to get involved. But really, it's about time because Safeway has so much to give. So do so many other big companies.'

Safeway said in a public statement: 'We are aware that the money we are contributing is a small gesture but raising levels of awareness is also crucial.'

Let's hope it's just the beginning.

Waterwise at work

Is there something you could do in your workplace? Don't just wait for Christmas to do your little bit of company charity work. Suggest a drought fundraiser to your boss. Maybe if employees raise a certain amount, the company will match it.

And don't stop there.

No matter what industry you're in, you can try this suggestion. Read the water meter at your workplace. By regularly reading your water meter you will be able to detect changes in usage patterns. A sharp increase in water used could indicate a leak in the system, which could be then rectified to prevent continued waste. Displaying meter readings in staffed areas may also be used as a motivator for staff to reduce water use. Consider setting water-saving goals and promote staff achievements when these goals are met.

WATER POLITICS

In the US city of Detroit, social services workers have removed children from more than 700 families because they cannot afford to buy water

The water shortage is a crisis, but it's not the only one we're facing. Though we don't hear as much about it, people around the world are also suffering the lack of water justice. Water should be a human right, but it is rapidly becoming a commodity. As water is privatised, the world's poorest people will suffer first.

A plastic problem

In many parts of the Third World, and particularly in South-East Asian countries, governments are starting to sign away their control over domestic water supplies by participating in international trade treaties. The World Bank has adopted a policy of water privatisation and full-cost water pricing.

Global multinational companies are buying up local water supplies. The sums they offer are often too tempting for governments to refuse. Then they bottle the water and sell it back to the locals – if they can afford it. The companies charge inflated prices and make huge profits.

Two hundred billion litres of water were produced in plastic bottles last year. Imagine how Third World countries that don't have the capacity to recycle deal with the mountain of plastic that piles up in their community each year.

In 2002, I travelled Vietnam and went to the beautiful world heritage area of Halong Bay. On the boat trip to one of the islands I noticed a sign that said 'Please place all rubbish in the bin.' Fantastic, I thought.

We arrived at our destination and I happened to be the last one off the boat. I walked on shore and turned back to see one of the boat hands throw all of the rubbish into the ocean! This image has stuck in my mind.

Education programs teaching people about recycling are vital, but can Third World countries afford them? I doubt this boat hand understood the ramifications of his actions. The generations before us acted in the same way, but until recently the products they were handling were organic, not manufactured petrochemicals with a 500-year half life.

If you travel in South-East Asia and closely examine the small print on your water bottle, you'll invariably see the mark of Coca-Cola. Coca-Cola provides 10 per cent of the world's total liquid intake. The company's ten-year plan is to reach 20 per cent. Imagine Vietnam in 2017 …

Marketing with Bling

Bottled water has become a commodity in developed countries, too, and the marketing experts roam free. I remember the very first brand of bottled water in Australia and now the market is flooded with labels. Each professes to be the purest and healthiest, while some campaigns claim that their product is better for you than water, with added this or that to help you replace this or that in your diet.

Water is now a fashion item, everyone carrying their own bottle to complement the mobile phone in their other hand. The type you drink is important. The shape of the bottle represents your character: sporty, smooth, rich, healthy ... So it doesn't surprise me that Hollywood has taken bottled water to new heights.

Writer-producer Kevin G Boyd has come up with a product that is designed for the rich and famous. Working in Hollywood, he noticed the correlation between the stars' status and the type of water they were carrying, so he came up with his own label: Bling H_2O. For people who don't watch American television, 'bling' is a word that describes the riches someone wears on their body for show (usually diamonds). It is the ultimate fashion statement. And speaking of statements, here is part of the mission statement from the Bling H_2O website:

'Our product is strategically positioned to target the expanding super-luxury consumer market.'

Paris Hilton is one of the product's main ambassadors and she even feeds it to her dog. Bling H_2O costs around $50 per 750 millilitre bottle. (And I am sure it's worth every penny.) As the website says, 'It's not for everyone, only those who bling.'

Bling H_2O won the gold medal at this year's Berkeley Springs International Water Tasting Festival. I wonder how people in Third World countries who have been conned into signing over their water rights would feel about that.

And don't think this could only happen in America. A Tasmanian man recently figured out that the world wants the water that falls from the clouds over Bass Strait. It's clean, safe, sexy and people are willing to pay. McFie's King Island Cloud Juice sells for around $US10 for a 750 millilitre glass. It's on the menu in chic Parisian water bars and high-end restaurants in Spain. McFie says he would like to see production grow to 500,000 bottles a year, up from his current output of 70,000 bottles of water a year.

Why do we need to carry water as an accessory item? Do we need to buy a new bottle of water each day? We have such high-quality of water in Australia! By buying less water in bottles and refilling them from the tap more often, we can reduce landfill and limit the power of these multinational companies. When buying a bottle of water, check who distributes the product. Are you really buying a bottle of good clean wholesome water, or are you buying Pepsi or Coke? You may be able to afford it, but not everyone in the world can!

What's it worth?

What price should we put on water? How can we guarantee that everyone in Australia – and around the world – will be able to afford water into the future?

Privatisation of water has one major benefit. When water is privatised, the price will rise and when we feel it in our hip pockets we will finally start to understand its real value. This will change the way we view and use it. Water prices in Australia are so low that we do not really value it enough.

I think we need a balance between public and private ownership, but we also need to ensure that everyone has access to it.

Some water activists suggest that the revenue raised from increased water rates should go back into the system. Share holders should receive fair repayments, and profits should be channelled back into our water infrastructure to clean and refill aquifers. Some of the earnings from water sales should be used to subsidise the provision of water to those who truly cannot afford it. The money should not go toward padding CEO salaries. If any policing is going to be done, it should be to ensure people are not suffering, and not to help the rich grow richer by denying those on low incomes a basic human right.

Water bandits

There's no Bling water and no Cloud Juice in drought-stricken areas in rural Australia. The big thing is water theft. We're starting to hear accounts of late-night robberies – not only from the local reservoirs, but from people's rainwater tanks as well.

'I came to the tap to get a glass of water and all I got was mud. I couldn't really believe it had happened,' Gundaroo farmer Lindy Hayman told journalist Rob Taylor late in 2006. In an article published online by News.com.au, Taylor reported that Hayman had lost 75,000 litres from her tank while she was out.

Kerry Wagstaff, also from Gundaroo, said that thieves emptied two 30,000-litre water tanks that provided the water for her house and vegetable garden. In nearby Bungendore water was stolen from village dams and tanks. Desperate times!

'Whiskey's for drinking ...'

Mark Twain once commented that in dry California, 'whiskey's for drinking and water's for fighting over.' In many parts of the world today, access to water is still so precious and so precarious that people have to fight for it, and the problems are only growing worse. That is why it is essential that we rally together. It is time to act as a community. If access to water must be restricted, we have to find a way to share this resource fairly.

The more we learn about the politics of water, the easier it will be. Two books I'd recommend are *Blue Gold: The Corporate Theft of the World's Water*, by Maude Barlow and Tony Clarke (Earthscan, 2002) and Tim Flannery's *The Weather Makers: The History & Future Impact of Climate Change* (Text Publishing, 2005).

If it seems the government is slow to cut red tape, if you are frustrated by the waste and mismanagement you see, start a lobby group, write letters, educate yourself and those around you. If all else fails, run for local government!

Finally, listen to your children – it is their generation who will find greater solutions.

Whiskey might be for drinking, but water's not for fighting over. Let's try to prove Mark Twain wrong ... at least about that second part!

EDUCATION

We all need to become Aquavists

I'm sure many of you remember that shocking moment in Year 8 biology class when we were informed that our body is more than 70 per cent water. My thought was that at any moment I might melt to the floor! It really impressed upon me the fact that all living things share that same need for water. It's an important message: we all need to become Aquavists. Education is the answer – but it has to be done right.

Blair Nancarrow is the director of the Australian Research Centre for Water in Society at CSIRO Land and Water. She says we need to have realistic expectations.

'Education as just an end in itself is not necessarily going to change behaviour. Too many education programs just basically explain what people need to do to save water. There's no acknowledgement of the fact that people might not *want* to do these things. There's no attempt to analyse behaviour to come up with better solutions. Nice pamphlets aren't going to help here.

'We need proper social analysis. Until you understand how and why people behave a certain way, you're not going to change anything.

'When I go out anywhere and tell people I work in water, the next thing they say is "Why don't we do such and such?" and all these great conservation ideas start flowing. People are passionate about water right now. This is a good opportunity to really keep the debate going. A lot of progress can probably be made.

'But the reality is this: there will be a limit as to when lifestyle flips in and the ability to save water in the household is going to stop.

'The situation at this stage is that people will not trade off. They want their showers. They want affordable shower heads that don't compromise their comfort. That's where technical solutions become very important.

'There are a lot of ways where we can fit lifestyle and technical fixes together. The work we're coming to now and over the next year is to look at exactly this. We're looking at design. Societal issues are very important. For example, getting the teenage child out of the shower without causing World War III is not worth it.

'We need to better understand what things people are more likely to do at a household level and how they fit their attitudes to sustainability into their busy lives. People who advocate a conservation effort don't use less water than the people who say "I've got a right to water and I've got to have it." We don't know if people decide to get a low-flow shower head so they can feel okay about using more on their roses or in their garden.'

Nigel Finney, the conservation guru behind the savewater! Alliance agrees. He knows how hard it is to change people's habits.

'If you've been taking twenty-minute showers all your life and I ask you to voluntarily drop down to four minutes, people are going to say "No way, I don't want to do it."'

Like Blair Nancarrow, Finney emphasises the importance of research into new technologies. 'People say, "Just give me a product to fix the problem; I don't want to change my habits." So we've got three-star-rated washers and dishwashers and tank-to-toilet water-recycling systems. A number of new innovations have come online. One is a device that goes underneath the bathroom wash basin and recirculates the water until it's warm, so you don't waste water by letting it run while it heats up.

'If you accept conventional wisdom that climate change has impacted on our water supply, you have to look at devices like solar hot water units and heat pumps which are very low energy users. Such innovations are important to the future of saving water. They are being encouraged through bodies like the Smart Water Fund, which provides $4 million a year to water conservation research projects.

'New air-to-water machines that pull moisture out of the air are commercially available now. The downside is that they use electricity to run. But they generate 25 to 45 litres a day, which is more than enough for consumption. People are scouting around for ways to make a solar-powered version of this machine. We're going to see a lot of waterwise technology coming into the market. Price is a bit of an issue but there are programs that will fund people who want to make the change.

'A lot of change still has to happen at the micro level. We're not all out there buying low-flow shower heads. We're not doing it.'

Nigel Finney is right: culture change only happens slowly. I'm optimistic, though, because it is happening. Households across the nation have cut their water consumption by as much as 20 per cent in response to rationing imposed by state governments.

The Australian Bureau of Statistics' water account showed deep cuts in urban and rural consumption of water between 2001 and 2006 in response to the prolonged drought.

Victorian households led the country with the biggest drop in consumption – 42,000 litres a year per household (a cut of 20 per cent)– as well as the lowest levels of household water consumption.

Households in south-eastern Australia have delivered average water cuts of at least 15 per cent, with more modest cuts in Western Australia, but increases in Tasmania, where no restrictions are in place.

Farmers have reduced their consumption by 19 per cent over the same period, with the value of output falling by only 6 per cent, reflecting increased efficiency and water being traded to higher value crops.

When the Australian Bureau of Statistics released this report, experts said it showed that most of the easy cuts had already been made. Increased water efficiency is no longer enough: we need a renewed focus on microeconomic reform of the national water market to deliver further improvements.

Personally I think the answer lies with the next generation of kids. They're the easiest ones to influence.

If and when the current drought breaks, climate change will still ensure that our future is a lot drier than the past. These kids will grow up conservationists – because they'll have to.

I remember a talk I gave to a group of seven- and eight-year-olds at a western districts primary school in Victoria. I asked the children how they thought they could save water and they came up with heaps of ideas. The first one was the old standard, of course: Don't leave the tap running while you brush your teeth. The young man who suggested it was a proud Aquavist and told me that his dad always left the tap running and that he'd told him on more than one occasion why it was a bad thing to do.

Be a teacher

When people talk about sustainability, they are talking about making sure resources like water are available for the next generation. The savewater! website has lots of good tips for teaching kids to be waterwise. If you have children of your own, or if you're a teacher, you could try some of the following activities:

- Rig up a system to measure rainfall in your local area.
- Carry out a water audit at home or at school.
- Visit the dam or treatment plant your water comes from and find out how it works.
- Debate water-related issues in your local area.
- Read newspaper articles about the water shortage together or discuss programs you see on television.
- Check out your local water authority's website.
- Investigate the role that water plays in various cultures.

- Find out about how access to water and water quality affect our lives and the lives of children in other countries.
- Paint pictures, sing songs or put on plays with a water conservation message.
- Write stories or poems about the importance of water in our lives.

The possibilities are endless!

**THE FUTURE:
OURS TO HOLD!**

Over 5 million houses are connected to mains water in Australia

Since Amy and I started writing this book, water and climate change have become hot topics, splashed across the front pages of the newspapers. People across the nation are starting to understand the issues – they're weighing on our minds.

I believe that most people are good and if they have trustworthy information, they'll do the right thing. In this book, I've passed on as many ideas about saving water around the home as possible. I hope you're inspired to put some of them into action – and then come up with a few of your own. If you do, drop me an email and, most importantly, tell your neighbours.

Whether you live in the city or the country, you can start at home by setting up a greywater system or harvesting rainwater. If it all seems too hard or too expensive, ask a local expert for some advice – and use your imagination. There's always a solution. My co-writer Amy and I are more than happy to answer any queries. You can reach us at info@thewaterbloke.com.au.

Talk to your friends, and get together in groups to push ideas to local councils. If you have a great idea for saving water in your local area, and support from your neighbours and friends, why not think about running for council yourself?

We need water to survive. We are made of water, and water keeps us alive. The crisis we face is severe, but we still have time. I believe deeply in our society: we will get our priorities in order. People are already moving, we are starting to make something happen. Be a leader: put those buckets in the shower, install a rain tank, switch to green energy. Once you know the basics, it's just a matter of habit, of practice and of thought. Australia's future is in our hands. Good luck!

CRAIG'S TOP TWENTY TIPS TO TAKE AWAY

- Keep showers under five minutes. Buy a shower timer.

- Install a water-saving shower head.

- Use buckets in the shower to catch 'warm-up' water.

- When you shave – whether it's your face, your legs or your armpits – fill up the hand basin or a small bucket to do the job, then rinse off in the shower.

- Consider a composting toilet – or at least install a dual flush system.

- Use rainwater to flush the toilet and fill the washing machine.

- Recycle your greywater onto the garden.

- Use drip irrigation in the garden or a hand trigger hose – whether you're recycling your water or not.

- Lay 75–100 millimetres of mulch on the garden to prevent evaporation. Straw mulches are the best.

- Use a pool cover.
- Change to green energy.
- Don't be afraid to call the authorities if you see a regular water waster.
- Get a water audit or assessment from your water provider or a private expert to find out where you can save.
- Check your plumbing for leaks regularly.
- Use a water-efficient dishwasher and washing machine.
- Plant new gardens in groupings according to the needs of particular plants.
- Fill the kettle just enough for a cuppa.
- Catch public transport whenever possible.
- Use recycled products (packaging and toilet paper).
- Share knowledge. Listen to your children!

PART 4
RESOURCES

WATER CONSERVATION PROGRAMS

The following sites focus on understanding how we use water and how to use it more efficiently.

National

CSIRO – Water Use and Reuse
www.clw.csiro.au/research/urban/reuse/index.html
The CSIRO's Water Use and Reuse team is working on cost-efficient ways to redesign Australia's water systems. The organisation has done pioneering studies into the environmental sustainability and economic viability of storage and reuse of water that would otherwise be wasted. For information about current projects, visit their website.

wwwtools – Online Learning Resources for Water Education
http://magazines.fasfind.com/wwwtools/m/37401.cfm
You could do a mini degree on water just by working through the wwwtools online magazine's water-related resources. The

site is loaded with info and has links to articles on everything from behaviour modification to water recycling and water management.

Australian Capital Territory

ACTEW – Water2WATER
www.actew.com.au/water2water
ACTEW provides water and wastewater services to the people of Canberra and surrounding areas, and has recently initiated the Water2WATER project, a plan to secure the ACT's water supply by purifying Canberra's used water and adding it to the Cotter reservoir. The site is packed with information on how this can be achieved and gives details of water-recycling schemes operating in cities around the world.

Think water, act water
www.thinkwater.act.gov.au
The 'Think water, act water' site provides information about the ACT government's rebate programs and services, and advice to help you make your home, garden, school or business more water-efficient. The site even features locals and their water-smart stories.

New South Wales

Sydney Water
www.sydneywater.com.au/EnsuringTheFuture/
WaterSchool/index.cfm

Sydney Water has created a water school by packing its website with engaging and easy-to-read resources on recycled water, water quality, saving water, wastewater and sewerage. It has fact sheets on the water supply system across Sydney, Illawarra and the Blue Mountains, rainfall trends and the history of Sydney's water supply. The site even has games to get kids involved.

Northern Territory

Power and Water

www.nt.gov.au/powerwater

Power and Water is the Northern Territory's water provider. Check its site for information on Darwin River Dam levels and information on using reclaimed water for irrigation schemes. The site also has news on the Water Reuse in the Alice project.

Queensland

Brisbane Institute

www.brisinst.org.au/resources/brisbane_institute_queensland_water.html

The Brisbane Institute is an organisation promoting informal learning through discussion and networking. It's worth checking out what members of this independent and innovative body have to say on the water issue.

Department of Natural Resources and Water – WaterWise gardening

www.nrw.qld.gov.au/water/saverscheme/waterwise_gardening.html

The Department of Natural Resources and Water in Queensland maintains a terrific site full of tips on waterwise gardening. Learn more about planning a water-efficient garden, watering your garden, mulch, maintaining a lawn and understanding your soil type.

South Australia

SA Water

www.sawater.com.au/SAWater/YourHome/SaveWaterInYourHome

The SA Water site lists top water-saving tips and up-to-date info on restrictions. You can also find out about waterwise appliances, rebates and home water audits. There are guides for charting your water use, suggestions on how to save water when caring for a pet, and useful contacts and links.

WaterCare

www.watercare.net

WaterCare is a South Australian government initiative that links programs and campaigns run by government agencies and local water boards that manage the state's water resources. It has information for everyone in the community: students, teachers, land managers and business and industry.

Tasmania

Department of Primary Industry and Water
www.dpiw.tas.gov.au
Tasmania's Department of Primary Industry and Water website has good tips on how to save water around the home, in the yard and garden and how to prevent water pollution.

Victoria

Department of Sustainability and Environment – Our Water Our Future
www.ourwater.vic.gov.au
Our Water Our Future is an initiative of Victoria's Department of Sustainability and Environment. Projects include the Schools Water Efficiency Program, which helps schools cut down on their water use, and 'Water – Learn It! Live It!', which aims to incorporate water conservation education into the state curriculum.

Melbourne Water
www.melbournewater.com.au
Melbourne Water has a range of online educational resources for primary, secondary and tertiary students. It also offers educational tours to various locations within Melbourne's water system and has a stormwater education trailer that comes to you.

Water Smart Home

www.museum.vic.gov.au/watersmarthome

Melbourne Museum has an interactive display inviting visitors to reduce water use in the home and garden. There's also a website with water-saving information, leading innovations in water-saving technologies and some inspiring examples.

Yarra Valley Water – Water School

www.yvw.com.au/waterschool

The Yarra Valley Water website features a 'Water School' section containing resources for primary and high school students as well as a section for teachers.

Western Australia

Water Corporation of Western Australia

www.watercorporation.com.au

The Water Corporation of Western Australia has a well-developed education website with teacher and student sections. The site covers all aspects of water education – not just conservation – to put it all into context.

Western Australia Department of Water

http://portal.water.wa.gov.au

Western Australia's Department of Water website has heaps of information about policy, tips on being waterwise and even careers in the water industry. You can also find out about the rebates offered on water-saving devices in the state.

BROADER CONSERVATION AND SUSTAINABILITY PROGRAMS

The following national organisations and programs focus on more than water conservation: they look at saving other resources such as gas and electricity and developing a sustainable society. Check out their websites for ideas about cleaning up our homes, workplaces and natural environment, safeguarding our native flora and fauna, investing ethically, halting climate change – and saying no to plastic bags.

Association of Societies for Growing Australian Plants (ASGAP)
http://asgap.org.au

Australian Bush Heritage
www.bushheritage.org.au

Australian Conservation Foundation
www.acfonline.org.au

Australian Research Institute in Education for Sustainability (ARIES)
www.aries.mq.edu.au

Australian SAM Sustainability Index (AuSSI)
www.aussi.net.au

Clean up Australia
www.cleanup.com.au

Climate Action Network Australia (CANA)
www.cana.net.au

CSIRO Sustainable Ecosystems
www.cse.csiro.au

Department of Environment and Water Resources: Ecologically Sustainable Development
www.environment.gov.au/esd

Eco Sustainable Developments
www.ecosustainable.com.au

Green Building Council of Australia
www.gbcaus.org

GreenPower
www.greenpower.com.au

No Plastic Bags
www.noplasticbags.org.au

Planet Ark Plastic Bag reduction campaign
www.planetark.com/plasticbags

Sustainable Business Practices
www.sbpractices.com

Work Energy Smart
www.energysmart.com.au

CRAIG'S RECOMMENDED PRODUCTS

Here are some great suppliers you may wish to contact for water-saving products.

Garden parts

Antelco
www.antelco.com
Irrigation parts.

Netafim
www.netafim.com.au
Drip line irrigation systems.

Reece plumbing and irrigation
www.reece.com.au
Fantastic advice for all water-saving solutions in and around the home.

Toro Australia Pty Ltd
www.toro.com.au
A full range of garden and irrigation parts.

Greywater

Aqua Reviva
www.aquareviva.com.au
Greywater treatment systems.

Ecocare
www.irrigationwarehouse.com.au
Greywater diversion systems, from sump pump to irrigation.

Lawnmowers

EnvirOmower
www.enviromower.com.au
Petrol-free lawnmowers.

Masport
www.masport.co.nz
Electric lawnmowers and hand mowers.

Rover Mowers
www.rovermowers.com.au
Hand mowers.

Victa
www.victa.com.au.
Electric mowers.

Mulches

Biogreen Limited
www.biogreen.info

Rainwater tanks

Nylex Water
www.nylex.com.au
A full range of water-saving products.

Polymaster Rainwater tanks
www.polymaster.com.au
Rain tanks and tank-to-toilet systems.

Reece plumbing and irrigation
www.reece.com.au
Great advice on water-saving in the home and garden.

Tankworld Rainwater
www.tankworld.com.au
For all rainwater harvesting solutions.

Waterwall Solutions
www.waterwall.com.au

Modular tanks, great for rainwater harvesting solutions in urban areas. Innovative design and friendly advice.

Redirecting thermostats

Everwater Australia
www.everwater.com.au
Everwater makes a redirecting thermostat unit known as the Chili Pepper, which recirculates water through mains pipes until it has heated up, eliminating waste.

Hydrotherm Australia
www.greenheat.com.au
Hydrotherm makes the GreenHeat Pump Kit, another redirecting thermostat.

Shower heads

Flexispray
www.flexispray.com.au
Low-flow shower specialists.

Interbath
www.interbath.com
Shower heads with a maximum flow rate of 9 litres a minute.

Tiara Showers
www.tiarashowers.com.au
Low-flow shower heads.

Water Wizz
www.waterwizz.com.au
Low-cost devices to fit in existing showers and taps, reducing water use by up to 50 per cent.

Shower timers

Ripple Products
www.rippleproducts.com

Soil enhancers/wetting agents

Debco SaturAid
www.debco.com.au

Fytogreen Australia
www.fytogreen.com.au

Toilets

Try local plumbing stores for dual-flush buttons to add to your existing toilet.

Caroma
www.caroma.com.au
Dual-flush toilets.

Clivus Multrum
www.clivusmultrum.com.au

Enviro-Fresh
www.enviro-fresh.com.au
Specialist in water-saving urinals.

Envirolet
www.envirolet.com
Composting toilets.

Nature Loo
www.nature-loo.com.au
Composting toilets.

Washing machines/dishwashers

Asko
www.asko.com.au

Miele
www.miele.com.au

Whirlpool
www.whirlpool.com.au

Waterless car wash

Meguiar's
www.meguiars.com.au

NoWet Waterless Car Clean
www.nowetcarclean.com.au

AMY'S RECOMMENDED PRODUCTS

Hair care

Okay, so you want to get into greywater. You want a luscious garden. You want great hair too – and you don't want your shampoo to screw it all up for you. When you're recycling shower water on your garden, your plants end up drinking your shampoo, conditioner and soap. Some products are easier on your blooms and won't leave you with a month of bad hair days. Check out www.neco.com.au. It's Australia's online eco superstore.

Go to the greywater personal care section and choose from the Alchemy line. The line has something special for all kinds of hair. If your hair is dry, use their Rice Aminos and Wheat Protein Intensive Moisture Shampoo. Those with oily or combination hair will love the Lemongrass Shampoo. It has six natural cleansers that gently dissolve oily scalp secretions without stripping the skin of its natural oils. The conditioners are equally divine.

Hand and body soap

You need to wash those hands, mister. Check out Rambilldeene Farm Sandalwood Liquid Soap, also available at www.neco.com.au. It comes in a handy pump pack and uses soybeans as its base ingredient. The soap is 100 per cent biodegradable and safe for greywater systems. The Rainforest Remedies bar soap is too, and it's only a few bucks.

Toilet paper

Merino's SAFE brand is probably the best known recycled toilet paper, but there are several other brands available now, including: Softex, Elite, Nature Soft, Tree Free, Naturale, Nice and Soft, Eco, and Ocean Soft. Generic brands such as Black & Gold, Payless and No Frills are often made from 100 per cent recycled paper, too, but they may not be unbleached.

Dishwashing detergent

Check out www.ecoathome.com.au for an amazing range of environmentally friendly dishwashing products. Bio-logic does dishwashing liquid in lavender, lemon lime and sweet orange for around $9. If you prefer to use a machine, check out Herbon Dishwasher Machine Powder. It contains only natural mineral biodegradable ingredients that leave no residue. A 1 kilogram package costs around $15.

Laundry detergent

Phosphate-free laundry detergent is easy to find. It'll be at your local grocery store. Check out Planet Ark. My personal favourite is Bio Zet Advanced Concentrate. All the ingredients are fully biodegradable. After washing, the enzymes break down naturally and are absorbed harmlessly back into nature. Bio Zet's packaging is also made from recycled paper.

You can find natural alternatives to bleach and harsh laundry detergents online, too. One site you might like to check out is www.naturallyclean.com.au. Naturally Clean makes a product called 'Snow', a high-strength formula for removing tough stains that is still gentle on the environment.

Lots of soaps and laundry detergents contain fresh-smelling eucalyptus that can hurt your plants. Make sure to avoid products with eucalyptus oil if you are recycling water for use on your garden.

Cleaning cloths

Microfibre cloths are great for picking up dust and dirt. Enjo is one of the best known brands. DYNA-WIPES are another great option. They're heavy-duty non-abrasive cleaning towels impregnated with a waterless hand cleaner. You can use them to remove grease from your stove and hard-to-remove marks from your counter tops. They'll also get rid of scuff marks on the floor, as well as paint and nail polish. They're worth a try. Order them online at www.dutchguard.com/p-dyna-wipes.html.

All-purpose cleaners

EnviroClean Vigor and Latis Intensive Cleaner are some of my favourite biodegradable, non-toxic, all-purpose cleaners that are great for kitchens. They're available online at www.energyandwatersolutions.com.au/cleaning/cleaning.htm.

WATER AUTHORITIES

Here is a state-by-state list of water and environmental authority contacts.

National

Department of the Environment and Water Resources
www.environment.gov.au
Phone: 02 6274 1111

Malcolm Turnbull
Minister for the Environment and Water Resources
www.malcolmturnbull.com.au
Email: malcolm.turnbull.MP@aph.gov.au
Phone: 02 9369 5221 or 02 9369 5225

Malcolm Turnbull is the current federal government's Minister for the Environment and Water Resources. His website has many useful links to information on water-related initiatives around Australia.

Anthony Albanese, MP
Shadow Minister for Water and Infrastructure
www.anthonyalbanese.com.au
Email: A.Albanese.MP@aph.gov.au
Phone: 02 9564 3588

Peter Garrett, MP
Shadow Minister for Climate Change, Environment and Heritage
www.petergarrett.com.au
Email: Peter.Garrett.MP@aph.gov.au
Phone: 02 9349 6007

Australian Capital Territory

Actew
www.actew.com.au
Email: waterconservation@actew.com.au
Rebates: www.thinkwater.act.gov.au
General inquiries: 02 6248 3111
After-hours: 13 14 93
Emergency: 13 11 93
Water Conservation Office: 02 6248 3131

Environment ACT
www.environment.act.gov.au
General inquiries: 13 22 81

New South Wales

Department of Environment and Conservation
www.epa.nsw.gov.au
Email: info@environment.nsw.gov.au
General inquiries: 13 15 55
Head office: 02 9995 5000

Sydney Water
www.sydneywater.com.au
Email: on.tap@sydneywater.com.au
General inquiries: 13 20 92
Emergency/faults: 13 20 90

Northern Territory

The Department of Natural Resources, Environment and the Arts
www.nreta.nt.gov.au
Email: epaboard.nreta@nt.gov.au
General inquiries: 08 8924 4139

Power and Water
www.powerwater.com.au
Email: customerservice@powerwater.com.au
General inquiries: 1800 245 092
Emergency: 1800 245 090

Queensland

Department of Natural Resources and Water
www.nrw.qld.gov.au
Email: enquiries@nrw.qld.gov.au
General inquiries: 13 13 04
New water infrastructure projects: 1800 243 585

Environmental Protection Agency
www.epa.qld.gov.au
Email: csc@epa.qld.gov.au
General inquiries: 07 3227 8185

SEQWater
www.seqwater.com.au
General inquiries: 07 3229 3399
Drought hotline: 1300 789 906

South Australia

The Department of Water, Land and Biodiversity Conservation
www.dwlbc.sa.gov.au
General inquiries: 08 8463 6800

Environment Protection Authority SA
www.epa.sa.gov.au
Email: epainfo@epa.sa.gov.au
General inquiries: 08 8204 2000
Complaints: 08 8204 2004

SA Water
www.sawater.com.au
Email: customerservice@sawater.com.au
General inquiries: 1300 650 950
Water restrictions hotline: 1800 130 952

Tasmania

Department of Primary Industries and Water
www.dpiw.tas.gov.au
General inquiries: 1300 368 550

Hobart Water
www.hobartwater.com.au
Email: hobartwater@hobartwater.com.au
General inquiries: 03 6233 6533
Emergency after-hours: 03 6261 2122

Victoria

Environment Protection Authority Victoria
www.epa.vic.gov.au
General inquiries: 03 9695 2722

Melbourne Water
www.melbournewater.com.au
General inquiries: 13 17 22
After-hours emergency/faults: 13 24 46

Our Water Our Future
Department of Sustainability and Environment
www.ourwater.vic.gov.au
Email: ourwater@dse.vic.gov.au
General inquiries: 13 61 86
Inquiries/complaints: 131 WATER or 13 19 28 37

Western Australia

Department of Water
www.water.wa.gov.au
Email: info@water.wa.gov.au
General inquiries: 08 6364 7600

Environmental Protection Authority of Western Australia
www.epa.wa.gov.au
Email: info@environment.wa.gov.au
General inquiries: 08 6364 6500

Water Corporation
www.watercorporation.com.au
Email: cust_centre@watercorporation.com.au
Water inquiries and complaints hotline: 13 10 39
Faults and emergencies: 13 13 75

GREEN POWER SUPPLIERS

ActewAGL (all states)
www.actewagl.com.au
Email: general.complaints@actewagl.com.au
General inquiries: 13 14 93
After-hours emergency: 13 11 93

Country Energy (ACT, NSW, QLD, SA, VIC)
www.countryenergy.com.au
Email: info@countryenergy.com.au
General inquiries: 13 23 56
After-hours: 13 20 80

Energex (QLD)
www.energex.com.au
General inquiries: 13 12 53
Emergencies: 13 19 62

Energy Australia (ACT, NSW, SA, VIC)
www.energyaustralia.com.au
Residential inquiries ACT & NSW: 13 15 35
All inquiries SA & VIC: 13 88 08

Ergon Energy (QLD)
www.ergon.com.au
Email: customerservice@ergon.com.au
General inquiries: 13 10 46
Faults: 13 22 96

Integral Energy (NSW, QLD)
www.integral.com.au
General inquiries: NSW: 13 10 02
Emergencies NSW: 13 10 03
General inquiries: QLD: 13 37 53
Emergencies QLD: 13 19 62

Origin energy (NSW, QLD, SA, VIC, WA)
www.originenergy.com.au
General inquiries: 13 24 63

Tru Energy (NSW, SA, VIC)
www.truenergy.com.au
Email: resolutions@truenergy.com.au
General inquiries: 13 34 66

Western Power (WA)
www.westernpower.com.au
General inquiries: 08 9326 4911
Faults/emergencies: 13 13 51

REBATES

This is a list of the major rebates available as this book goes to print. For updates, contact your water provider or check my website: www.thewaterbloke.com.au.

I've tried to include as much information as I can, but it's always best to speak to your water provider or state government authority for full details *before* you spend your money. You need to be sure you're eligible under the rules that apply in your area. For example, you might not be entitled to a rebate on a rainwater tank if tanks are mandatory in your suburb, or it may be necessary to have a licensed plumber install an aerator or set up your greywater system if you want to claim some of the purchase and installation costs back.

It's always worth calling your local council to check if you are eligible for any additional offers. These are sometimes quite generous. In Brisbane, residents can install a water tank for next to nothing, once they've added the state and local government rebates together!

Australian Capital Territory: Think water, act water

Product or service	Rebate	Conditions
Rain tanks	$550 (2000–3999 L) $700 (4000–8999 L) $800 (9000+ L)	The tank must have an internal plumbing connection.
Connecting an existing rain tank to a toilet or washing machine	$400	—
Replacing a single-flush toilet with a dual-flush	$100	You must book a WaterSmart Homes service for a fee of $30.
WaterSmart Homes service	—	For $30, a plumber will visit your home and install water-saving devices such as flow restrictors and a low-flow shower head. One shower head is included in the $30 fee. A second shower head costs $22. Participants can apply for a $100 rebate to replace a single-flush toilet with a dual-flush.
GardenSmart service	—	For $30, a horticulturist will visit your garden to assess your needs, then give you practical water-saving advice. (This kind of service would normally cost around $130.) Participants can apply for a rebate of up to $50 when they buy selected water-saving products.
Approved products	Up to $50	You must book a WaterSmart Homes service visit for a fee of $30.

Source: www.thinkwater.act.gov.au/tuneup_rebates.shtml

New South Wales: Sydney Water

Product or service	Rebate	Conditions
Rain tanks	$150 (2000–3999 L) $400 (4000–6999 L) $500 (7000+ L)	An additional $300 rebate is available if a toilet and/or washing machine is connected to the tank.
Washing machines	$150	Minimum rating: four stars.
WaterFix service	—	For $22, a plumber will visit your home to install water-saving devices, including a low-flow shower head and aerators or flow restrictors. They will also adjust suitable single-flush toilets and repair minor indoor and outdoor leaks. (This service would usually cost around $180).
Love Your Garden service	—	A horticulturist will visit your garden to assess your watering needs, then give you water-saving advice and tools tailored to your garden, such as rain gauges or tap timers, along with vouchers for products and services. An assessment on a block 1200m² or less is $33. For a block over 1200m² it is $55.

Source: www.sydneywater.com.au

South Australia: SA Water

Product or service	Rebate	Conditions
Connecting a rain tank to a toilet, washing machine or other internal fixture	$400 (1000+ L)	The rebate is offered on the plumbing work rather than on the tank itself. The rainwater tank must be connected to at least one internal fixture.
Flow restrictors Low-flow shower heads Tap timers	$10 per item up to a maximum of $50 OR $20 per item up to a maximum of $100 (SA Water and/or Family and Youth Services concession cardholders)	Customers need a receipt as proof of purchase of an approved water-saving device. For flow restrictors, customers must provide proof of installation by a licensed plumber. A maximum of one claim per household.

Source: www.sawater.com.au/SAWater/YourHome/SaveWaterInYourHome/Rebates.htm

Tasmania: Hobart City Council

Product or service	Rebate	Conditions
Rain tanks connected to a toilet	$220 (600+ L)	The tank should be specifically designed for flushing the toilet.
Rain tanks for watering the garden	$170 (1600+ L)	
Washing machines	Up to $100	Check with Hobart City Council for current rules on minimum ratings.
Dishwashers	Up to $100	Check with Hobart City Council for current rules on minimum ratings.
Replacing a single-flush toilet with a dual-flush	$40	
Low-flow shower heads	$10	
Water audits	$20	The audit must be performed by an accredited GreenCity Service provider (or equivalent).
Replacing or repairing leaking fixtures	$25	A plumber must declare that no leaking fixtures or appliances remain and that at least one leak was fixed.

Source: www.hobartcity.com.au

Victoria: Water Smart Gardens and Homes Rebate Scheme

Product or service	Rebate	Conditions
Rain tank	$150 (600+ L)	An additional $150 rebate is available if a toilet is connected to the tank.
Large rain tank connected to a toilet and/or washing machine	$500 (2000–4999 L)	
Large rain tank connected to a toilet or washing machine	$900 (5000+ L)	
Large rain tank connected to a toilet and washing machine	$1000 (5000+ L)	
Greywater systems	$500	
Dual-flush toilet	$50	
Low-flow shower head	$10	
Water conservation home audit	Up to $50	The audit must be carried out by a qualified professional.
Approved products	$30	To be eligible for the rebate, you must spend $100 on approved waterwise garden products.

Source: www.ourwater.vic.gov.au

Northern Territory: Waterwise Program

Product or service	Rebate	Conditions
Flow restrictors Low-flow shower heads Tap timers Trigger nozzles	$50	Rebates are available via participating retailers in Central Australia. In Alice Springs, these are Big O Mitre 10, Home Timber and Hardware, Reece Plumbing Centres, Taps, Tubs and Tiles and The Watershed. In Tennant Creek, BJ Trading and True Value Hardware are participating.

Source: www.nt.gov.au/nreta/naturalresources/water/waterwise

Queensland: Home Water Wise Rebate Scheme and the Home Garden Water Wise Rebate Scheme

Product or service	Rebate	Conditions
Rain tanks	Up to $1000 towards purchase and installation	No minimum size. If the tank is connected to inside fixtures, it must be installed by a plumber licensed in Queensland.
Greywater systems (above-ground)	50% of purchase price and up to $200 towards installation	All necessary local council approvals must be obtained prior to applying for the rebate. A maximum of one claim per household.
Greywater systems (below-ground)	$500 towards purchase and installation	The rebate is available only for greywater systems approved by the applicable local council. A maximum of one claim per household.

Washing machines	Up to $200	Minimum rating: four-stars. A maximum of one claim per household.
Replacing a single-flush toilet with a dual-flush	Up to $150 towards purchase and installation	A maximum of two claims per household.
Low-flow shower heads	$30 or 50% of purchase price (whichever is the lesser amount)	Minimum rating: three stars. A maximum of three claims per household.
Swimming pool covers and/or rollers	Up to $200 towards purchase and installation	Minimum size: 25 square metres. Minimum warranty: five years. The pool cover must be endorsed under the Smart Approved Watermark Scheme.
Approved water-saving garden products	50% of the total purchase cost, up to a maximum of $50	Advice must be sought from local nurseries to determine which plants are best suited to the local area. A maximum of one claim per property.

Source: www.nrw.qld.gov.au/water/saverscheme/rebate_schemes.html

Note: Brisbane City Council, the Gold Coast Council and Toowoomba City Council all offer generous rebates on water-saving products and services, so if you're a resident in one of these areas, be sure to contact your local council to find out if you are eligible.

Western Australia: WA Water Corporation

Product or service	Rebate	Conditions
Rain tanks, 600+ L	$50	
Rain tanks, 2000+ L	Lesser of $500 or up to 50%	The rain tank must be connected to a toilet and/or washing machine.
Greywater systems or aerobic treatment units	Lesser of $500 or 50% of the purchase/installation cost	Maximum of one claim per household.
Washing machines	$150	Minimum rating: four stars.
Low-flow shower heads	$10	Maximum of two claims per household.
Flow restrictor	$20	Minimum rating: three-stars.
Garden assessments	$30	The assessment must be performed by an endorsed Water Corporation Waterwise Garden Centre. (Normally costs around $65.)
Garden bores	The lesser of $300 or 50% of the installation cost	
Subsurface irrigation pipework (30 m roll)	$10	Pipework must be endorsed under the Smart Approved Watermark Scheme.
Rain sensors	$20	The rain sensor must be endorsed under the Smart Approved Watermark Scheme. Maximum of one claim per household.
Tap timers (automatic)	$10	Only specific timers are approved.
Pool covers	The lesser of $100 or 50% of the total cost	The cover must be endorsed under the Smart Approved Watermark Scheme. Minimum warranty: eight years.
Wetting agents	$10	Only specific wetting agents are approved for a rebate. Maximum of two wetting agent purchases per household per year.

Source: http://portal.water.wa.gov.au/portal/page/portal/WiseWaterUse/WaterwiseRebates

ACKNOWLEDGEMENTS

Craig

Thanks to the Jones Family for their wonderful spirit.

Thank you to my family for the great support: Mum and Dad, Nigel and Cara and Nana Madden. You have stood by me through all my excessive shifts.

To Antoni Jach, for your incredible belief in me and for your mentorship.

To Kerensa for sharing a beautiful journey I shall never forget.

I would also like to thank Dan the Ideas Man and all the 2006 RMIT masters writing students.

To Ben and Vixy, Jimmy, Damo, Luke, Feruu, Brian, Fiona, Trent, Camo and Betty and other friends and family who put up with nothing but water talk throughout the duration.

Very special thanks to Amy and Robby.

Amy

This book would not have been written if it weren't for the amazing writer and teacher, Antoni Jach. What a source of inspiration you are!

Thank you Jeanne Ryckmans for taking a chance on us.

Robby Jennings, I literally wouldn't be here without you. I would be lost without you.

RMIT creative writing students, your insight and ideas kept us going.

Kim, Ted and Sandy Carmichael, I miss you. Thanks for all your support.

Yvonne and Barry, thank you for welcoming me into your truly waterwise home.

Thanks to Charisse, Jerrod, Suzi, Daan, Danny, Rhiannon, Ayesha, Sam, Jane, Briony, Angela and Lee; you too endured water talk that probably seemed like it would never end.

Craig, cheers to the future, you've helped make it bright.